MW00761031

Michael McIntosh

# WILD THINGS

Michael McIntosh

# WILD THINGS

With Illustrations
from
WILDLIFE ART

Countrysport Press
New Albany, Ohio

This edition of *Wild Things* was printed by Jostens Book Manufacturing, State College, Pennsylvania. The book was designed by Angela Saxon of Saxon Design, Traverse City, Michigan. The text is set in Berling, an old-style roman typeface created by Karl-Erik Forsberg in 1951.



First Edition
10 9 8 7 6 5 4 3 2 1

Published by Countrysport Press
15 South High Street, New Albany, Ohio 43054-0166



Printed in the United States of America

ISBN 0-924357-57-6 (trade ed.)
ISBN 0-924357-58-4 (limited ed.)

Library of Congress Cataloging-in-Publication Data

McIntosh, Michael.
Wild things / Michael McIntosh ; with illustrations from Wildlife art. — 1st ed.
p. cm.
Compilation of articles originally written starting in 1989 and published in Wildlife art news.
1. Zoology—North America. 2. Game and game-birds—North America. I. Title.
QL151.M39 1996
591.97—dc20 96-5226
CIP

*To Charlie and Libby,*
*who showed me the ways of looking*

*To John,*
*who taught by example*
*how to write what I see*

*And to the wild things*
*we've loved so long and well*

# Contents

# Foreword

Ever since I was a young boy, I've enjoyed the outdoors. It's a fascination I inherited from my father. He enjoyed the North Shore of Lake Superior as a youngster, and his gift to the family, a cabin on the shore of that great lake, was heavenly for me. In my early childhood I had severe bouts of hay fever. My seeming misfortune was however good fortune, as I never went to school until the first frost came to Minnesota. This gave me a chance to spend time in the autumn woods observing deer, partridge, and other critters. And so I'm very grateful for the opportunity my parents gave me to enjoy wild things.

In my boyhood years, I was also very fortunate to have had a good teacher in our caretaker, Roy Hangartner, who maintained our home "up north." He instilled in me the urge to learn more about the animals he loved and became a great source of encouragement, cultivating my curiosity about the woodlands and wildlife. He willingly shared his knowledge and experiences with me and my family, and for that, I am very thankful.

My lifelong fascination with nature is what initially inspired the concept of doing an ongoing feature about wild animals. During my early years at *Wildlife Art*, a magazine I started in 1982, I wanted to provide nature information to readers both young and old about animals I had experienced under Roy's tutelage and through my parents' encouragement. I thought a magazine on wildlife art should also include information about the animals the artists chose to depict, because if we viewed wildlife art without knowing anything about the animals, a disparity between the artist's rendering and our appreciation would result.

Initially, I tried to write some of the first columns about animals in 1983, when *Wildlife Art* was in its infancy. But it wasn't until 1988, when I met Michael McIntosh, that my dream to include stories on the animals themselves came true. Michael, who was very excited about the project, also was well equipped to write about animals, as he had worked for years as a writer and editor with the Missouri Department of Conservation. Personally, I think it was in his blood to produce these stories. From that initial discussion, "Wild Things" (formerly "Along the Wilderness Trail") grew into one of our most-read features in *Wildlife Art*.

In the years since our first story on bears in March 1989 it has been a wonderful adventure. Every year Michael, editor Rebecca Hakala Rowland, and I pick a new group of subjects to cover in the year ahead. Michael then goes crazy doing the research and writing the fascinating stories, many of which you will be enjoying in this volume. His writing, which is enhanced by a wide breadth of personal experience, conveys the charming and often fascinating qualities of the animals that share this planet with us.

Seeing how well these features have been received, it was only natural that someone should want to make a book of them. So I was thrilled when Countrysport Press first approached us about pulling together many of Michael's essays to make

*Wild Things.* For me it is another dream come true. I'm hoping that this book can be readily distributed to schools as well as to those who want to learn more about everything from pine martens and marine mammals to bison and cardinals.

The artwork accompanying each story in this book has been gleaned from artwork submitted by the many artists who have participated in "remarque contests" sponsored by the magazine. A remarque is an original drawing done on the margin of a limited edition print. Special thanks to the artists who allowed us to use them.

I hope *Wild Things* provides hours of reading enjoyment for you. Maybe our friends at Countrysport will publish another edition some day that will include animals of the world.

Robert J. Koenke
Publisher/Editor in Chief
*Wildlife Art*

*Preface*

# A Life in the Woods

When I was very young I dreamed of someday writing a book.

Now I'm no longer so young, and I've written eight books and edited several more. They've all been on subjects dear to my heart—fine guns, hunting, sporting and wildlife art. They've made a bit of money and helped establish a reputation and a career in this odd, wonderful profession that chose me long ago.

This book is different.

This is the book I dreamed of writing before I was even old enough to read. This book is about the keystone that holds the subjects of all the others together.

I love guns and hunting and art for their own sake, always have and always will. They aren't particularly diverse as subjects go, but none of them represents the common thread, the central sinew that makes it all make sense.

This book does, for that one unifying element is an endless fascination and a deeply felt passion for nature in general and wild animals in particular.

The feeling has been with me from the beginning, woven inextricably through all my memories. I had the good fortune to spend my childhood in a time and place where the natural world loomed large and close. The woods literally began where the back yard ended, and they were playground and classroom and sanctuary all in one.

Of the books my parents and grandparents read to me, none was so enchanting as *The Little Pond in the Woods*. (Browsing in a bookstore not long ago, I was pleasantly surprised to find it still in print after all these years; if you have occasion to buy a gift for a child and are inclined toward books, I recommend it highly.) The great Disney nature films of the 1950s had much the same effect; I sat spellbound through so many Saturday-matinee showings of *The Living Desert* and *The Vanishing Prairie* and others that I had them almost memorized.

Nature simply was an essential presence in my life. I was aware that there was an abstract, academic side to it, but I felt no urgency for doing anything about it—which may be why I didn't take a single biology or zoology course during high school or college. Instead I chose to slake my endless curiosity by spending as much time as I could in the woods and fields hunting, fishing, prowling around, absorbing by osmosis such natural lore as came my way. The reading I did on my own was entirely unstructured and focused solely on whatever captured my fancy most strongly. Consequently I learned a great deal about birds and mammals and insects, somewhat less about reptiles and fish, almost nothing of botany and dendrology.

It all began to change in the mid-1970s. As a lark and as a break from the discipline of the scholarly writing I was doing as a professor of literature, I began writing magazine stories about hunting and shotguns—and was astonished to find that I could sell them readily. It was a revelation that changed my life.

I left teaching not long after and hired on with the Missouri Department of Conservation as a writer and senior editor. That, too, was a revelation. Being immersed in wildlife and nature as the raw material of a full-time job proved more thoroughly satisfying than anything I'd done before. It felt like coming home.

It was the beginning of an intensive education as well. My job was to translate technical scientific data into layman's language, to present complex material in forms both clear and entertaining. In working with the professional biologists I learned the lingo and the concepts of their science, learned to understand their literature, and learned how their great body of knowledge is organized. It was an invaluable experience both for the intellectual growth and for the opportunity to experiment with an approach to natural-history writing that is at once factual and anecdotal, realistic and yet laced with the emotion and sense of wonder that wild things have always stirred in me.

By the late 1980s I was on my own as a full-time freelancer, working almost exclusively with fine guns and bird hunting and feeling some sense of loss at not having the pleasures of wildlife research and writing. And then in the fall of 1988, as I was browsing the current issue of *Wildlife Art News*, it occurred to me that a magazine devoted to wildlife art might have some interest in stories about animals. I wrote to Bob Koenke, founder and publisher, and suggested an ongoing series of feature articles, each one focused on a specific animal, where it lives, what it eats and what eats it, how it makes its living and reproduces, who its relatives are, how it has affected mankind and how mankind has affected it, all to be illustrated with the editors' choices among the best animal artists of the past and present.

As it turned out, Bob was way ahead of me, already mulling exactly the same idea; he wrote back right away to say that the first piece was scheduled two issues hence and that if I cared to come by the editorial offices on the way to or from my annual grouse-hunting excursion to northern Minnesota we could decide on the first year's subject matter. I did and we did and that's how these stories came to be. The first one, about bears, appeared in the March/April issue of 1989, and there's been one in every issue since. For a long time they appeared under the heading of "Along the Wilderness Trail." When the magazine was redesigned last year and its title shortened to *Wildlife Art*, my department's title changed, too; now it's simply "Wild Things."

Fitting editorial material into limited space is a perennial problem for those who publish magazines, especially when their writers are as long-winded as I sometimes am, so a few of these pieces had to be trimmed for *Wildlife Art*. As books offer more latitude in that regard, what you'll find here are the original, longer versions. And in all of them I've done some minor re-editing, tinkering, and updating.

In the process of selecting what I consider the best of the series, and for the first time thinking of these stories as pieces of a whole rather than individual projects, I notice some things that seem worthy of comment.

I am both a hunter and a champion of animals and make no apologies on either account. All the good hunters I know feel the same way. Those who see hunting as merely killing have no idea what hunting really is about; those who would hate any creature that might chance to eat a game animal are equally ignorant. That one faction might call me a murderer and the other a bunny hugger troubles me not in the least. Anyone who really understands how nature works knows that to nature species are more important than individual animals, and knows as well that predators have as much moral right to their place on earth as prey.

If you wish to define me in a word, call me a conservationist. *Conservation* means "wise use," and it sums up my view of how we ought to treat the world we live in and the creatures that share it with us. It means using in reasonable ways those resources that are abundant, and protecting those that are in jeopardy of being used up.

I notice, too, that I tend to be hard on my own species for the way we've treated wild animals. No apologies for that either.

If you want to send me into the gloomiest depths of pessimism, just get me started on the notion that we're likely to be the only species of animal that ever destroyed an entire planet. And if that happens we'll have more to answer for than the author of Revelation ever dreamed.

On a happier note I see that I tend to talk about every animal as if it were my particular favorite. A lot of people who love wild things do the same, and it only means that there's something interesting and admirable about any animal you care to name. Years ago, for instance, I wrote a piece on ticks. I have absolutely no affection for the little bloodsuckers, especially during the time of year when picking them off my dog is a daily chore. On the other hand, their biology and physical adaptations are fascinating; I simply can't help admiring an animal whose heat sensors are sophisticated enough to recognize me as a warm-blooded animal at twenty feet or better, or that's able to live two years without eating.

I see as well that I frequently use Missouri as an example of how man and wildlife have interacted. The reasons are simple. I've lived in Missouri for twenty-five years now, and although I spend a lot of time traveling to hunt, fish, and otherwise ramble the wilder parts of the world, home is where my touch with nature is most intimate. Missouri is also about as diverse in both landscape and fauna as any place you'll find, and for more than fifty years it's been the undisputed leader in wildlife conservation and management nationwide.

That I have been able to write these stories at all bespeaks a debt I owe to a certain few people who have influenced me in particular ways. My father went out of his way to see that I was exposed to nature and wild things almost from the time I was an infant, and neither he nor my mother ever stinted in encouraging me to write.

Charles C. Ayres Jr.—Mr. Ayres, as he appears in these chapters—was the first naturalist I ever knew whose knowledge was highly technical and whose enthusiasm was infectious.

Charles and Elizabeth Schwartz offered those same qualities in even greater proportion. Artist, biologist, photographer, filmmaker, and perennial teenager, Charlie may have been the most broadly knowledgeable and widely talented naturalist who ever lived. Together, he and Libby—who is a first-rate writer, biologist, editor, and researcher in her own right—knew more, accomplished more in aid of wildlife, and had more fun doing so, than anyone I know. They were there, in the thick of it, when the modern science of natural history and wildlife management was born. Charlie, incidentally, illustrated the original edition of *A Sand County Almanac.*

Over the years that we've been friends, they taught me more than they ever knew. When Charlie died, four years ago, we lost one of the real ones. I miss him and his boundless enthusiasm.

Libby once flattered me beyond words by calling me her favorite outdoors writer. She's not one to bandy such praise lightly, but I've always wondered if she hadn't for a moment forgotten a mutual friend of ours who is *my* favorite outdoors writer.

It isn't often that any realm of science produces a truly fine wordsmith. The customs of scientific language don't encourage it, and in fact, scientific language is the antithesis of evocative, poetic modes of expression. In the field of natural history, the best work often has come from people who are writers first and scientists second, or not at all. John McPhee, for instance, proves the old dictum that there are no dull subjects, only dull writers, and his occasional forays into natural history are invariably gems. Still, there have been a few trained professional naturalists who possessed the gift. Ernest Thompson Seton was one. Aldo Leopold and Rachel Carson had it, too. So did Loren Eiseley.

But to my mind John Madson is the finest of them all, the John Madson I mention and quote fairly often in these stories. To put it simply, in natural-history writing John was my hero—and still is, even though he, too, is now gone. For even longer than the twenty years we were friends his work has been the example and the ideal I turn to when I need reminding how the job ought to be done, how to combine sound, thorough research with genuine, unsentimental affection, and how to render it all in thoughtful, cleanly crafted prose.

That these pieces have appeared in print owes thanks to some other colleagues and friends: To Bob Koenke, Rebecca Rowland, Kay Hong, and the rest of the *Wildlife Art* staff, and to Jim Rutherford, Art DeLaurier Jr., and Shelley Koop, of Countrysport Press—all for believing in me and my work, for providing the opportunity to write about bits and pieces of my life in the woods and the insights and delights I find there.

Michael McIntosh
Copper Creek Farm
Camdenton, Missouri

# Some Mildly Technical Stuff

Although this is by no means a scientific book, the study of living things is a science. We call it biology. Biology in turn comprises a number of specialized branches, among them zoology, which is concerned with the nature of animals. Since animals are broadly defined as any living organism that isn't a plant, fungus, or one of the blue-green algae, zoology itself contains certain subdisciplines—among them entomology, the study of insects; herpetology, reptiles and amphibians; ichthyology, fishes; mammalogy, mammals; and ornithology, birds.

For the most part, just knowing this much is enough to see you safely through the chapters to come. What few specialized scientific terms I've used are defined as they occur. The notable exceptions are some of the language and concepts of taxonomy, and taking a few minutes to talk about them now may spare you some confusion later. It's interesting stuff, besides.

Taxonomy is the systematic classification of plants and animals according to their natural relationships and similarities. It is therefore the branch of biology that concerns itself with names.

Humans are compulsive name-givers. We don't seem to feel comfortable with anything, from a whale to a virus, unless we can assign it an identity, call it by name. To have a name is to have a place in the order of things—and attempting to perceive order is yet another human need.

We're particularly enthusiastic about naming the animals and plants with which we share the earth, and for the most part we're satisfied with straightforward, everyday words—bluebird and grasshopper, sycamore and catfish, rabbit and violet. Such common names serve perfectly well for most of us, but they won't do for science. Common names are imprecise. Most plants and animals are known by several, some by a dozen or more. There are at least four different birds called "pheasant" in different parts of the country. When I was a kid, one of my fishing partners was an old man who called almost everything he pulled out of the water a "perch." Biologists would have a tough time sharing information without some way of making sure they're all talking about the same thing. Language differences also pose problems.

The scientific community solves all this by giving each life form a particular, unique name by which it is known worldwide, a name composed from Latin or classical Greek, which are the nearest we have to universal languages.

Scientific names are often long and full of vowels, bizarre-sounding and arcane. Even the words used to describe them seem esoteric: because the science of classification and naming is called taxonomy, an individual name is a taxon, taxa in the plural.

But these names serve their purpose well, and besides distinguishing a particular life form from all others they imply the whole system of classification that has evolved as a means of organizing what we know—or think we know—about the nature of nature.

◆

In order to make sense out of the vast number of life forms on earth, natural scientists realized ages ago that they needed some system for grouping things according to similarities and differences. Aristotle made the first attempts at classifying animals nearly 2,300 years ago—though his perceptions often were influenced more by folklore than by fact. In the seventeenth century, English naturalist John Ray first applied similar principles of classification to both animals and plants, based on physical structure. Ray also offered what apparently is the earliest definition of *species*—generally the smallest unit into which life forms are grouped. A species, as Ray described it and as the term still is used, is a group whose members interbreed to produce offspring like themselves, and who are more like each other than like any members of other, similar groups.

A word about the word itself. *Species* is pronounced SPEE-shees. *Species* is both singular and plural; one species or many species, it's all the same. *Specie,* which you'll sometimes hear misused as the supposed singular of *species,* is a noun that means "money in coin" or "a response in kind."

Important as John Ray's contributions were, it was Karl von Linne, an eighteenth-century Swedish naturalist, who in 1753 established the principles of classification that are the foundation of the system used today. Linne's intention literally was to name every living thing, and he so immersed himself in the task that he Latinized even his own name and was thereafter known to history and science as Carolus Linnaeus.

In a burst of youthful enthusiasm, Linne and his schoolmate Pehr Artedi divided Creation between them, Linne to name all the plants and Artedi the animals. Artedi's dedication got short-circuited some time later when, making his way home one night from an Amsterdam tavern, he fell into a canal and drowned. Linne, who was either a teetotaler or more careful, lived to the age of seventy-one, by which time he had devised names for hundreds of plants and animals, a great many of which are accepted still.

◆

Under the current system, of which Linne is considered the father if not the actual author, life forms are divided into a system of progressively more specific groups. It begins with two: the kingdoms plant and animal. There is considerable sentiment at present to think of the fungi and the blue-green algae as belonging to kingdoms of their own, but this is not yet universally accepted.

The kingdoms are subdivided into ever smaller, more specific categories wherein the individual members show progressively closer relationships to one another. These make up a multitude of sub-units—subkingdoms, divisions, phyla, subphyla, superclasses, classes, subclasses, infraclasses, orders, suborders, superfamilies, families, subfamilies, genera, and species. Species are sometimes further reduced to subspecies. Each group bears a Latin or Greek name that describes a key characteristic of its members.

For instance, the coyote belongs to the Animal Kingdom and to the following:

> The subkingdom Metazoa—animals whose bodies comprise many different types of cells that make up organs;
>
> The division Deuterostomia—animals whose mouths develop in a certain way during the embryonic stage;
>
> The phylum Chordata—animals with a primitive sort of backbone;
>
> The subphylum Vertebrata—animals having backbones composed of vertebrae and a braincase, or cranium;
>
> The superclass Gnathostomata—animals with true jaws;
>
> The class Mammalia—animals that feed their young on milk;
>
> The subclass Theria—mammals whose young develop for some time within the female reproductive apparatus;
>
> The infraclass Eutheria—mammals that are nourished by a placenta before birth;

The order Carnivora—flesh-eating mammals;

The suborder Arctoidea—the dog, weasel, bear, and raccoon families;

The family Canidae—the dogs, wolves, coyotes, foxes, and jackals;

The genus *Canis*—the dogs, wolves, and coyotes.

These various groupings function something like a family tree that shows who the coyote's relatives are, in descending order from the most distant to the nearest, and like every good family tree, it ultimately arrives at individuals—or in the case of animals, individual species. To a mammalogist, the coyote is *Canis latrans.*

Although the species is for practical purposes the smallest taxonomic category, zoologists often further divide species into subspecies. Subspecies classification is used to describe certain populations of a particular species that consistently show some notable difference from other populations. These usually have to do with coloration or marking, and the populations usually are separated geographically. Individual animals of different subspecies will freely interbreed wherever their ranges overlap. Taxonomists who are inclined to recognize a lot of subspecies are colloquially known as splitters; those inclined the opposite way are called lumpers.

The names of species, both animal and plant, are made up of two words—the generic or genus name and the specific name, or epithet. By custom, the first is capitalized, the second is not, and both are rendered in italics. Also by custom, the generic name may be abbreviated as an initial in subsequent references, as in "*Canis latrans* is closely related to C. *lupus,* the wolf."

Subspecies names are made up of three words but are otherwise treated the same as species names. The two subspecies of coyotes, then, are *Canis latrans thamnos* and C. *l. frustror.*

For the simplest handle on how all this works, think of scientific names as the equivalent of human names listed last-name first. To identify a coyote as *Canis latrans* is the same as identifying a person as Smith, John. Subspecies names are like distinguishing one John Smith from another by using middle names: Smith, John David, and Smith, John Thomas.

Traditionally, the person who first describes a previously unknown animal or plant is entitled to name it, and his own name, either in full or in abbreviated form, then becomes part of the official scientific name. Since Thomas Say, an early-nineteenth-century American naturalist, first described and named the coyote, the animal's full name technically is *Canis latrans* Say.

I don't get quite that technical in these stories, but if you happen to consult a field guide or other reference that does, the way the full name is listed will tell you something. For a hundred years or more, the science of taxonomy has been as much concerned with reclassification as with classification. During that time, our knowledge and understanding of life forms has expanded prodigiously, and this has led to a considerable amount of shuffling, reclassifying, and renaming. So, if you run across a taxon in which the describer's name appears in parentheses, this tells you that the animal originally was assigned to a genus different from the one it's currently placed in. The raccoon, for instance, is *Procyon lotor* (Linnaeus)—which means that Karl von Linne named it but placed it in some genus other than *Procyon*. (Actually, he thought it was a species of bear and assigned it to the genus *Ursus*, but you'll find all that in the raccoon chapter.)

The most interesting thing about scientific names is that they mean something. They can be translated, and the result often describes the animal better than its common name. In Latin, *canis* is "dog," and *latrans* is "one that barks." Combined, they identify a coyote as "barking dog," which is putting it mildly.

Names may derive from all kinds of sources. Some are Latinized versions of people's names, either the describer's or that of someone he wishes to honor. This is especially common in herpetology and botany. Epithets often are derived from the location where the original, or "type," specimen was collected. *Colinus virginianus*, "Virginia quail," indicates that the particular bobwhite from which the species was defined was taken in Virginia. *Odocoilus virginianus* means the same is true of white-tailed deer.

The most interesting names come from some characteristic of the animal itself—its appearance, habitat, behavior, voice, even its preferred foods. *Anas acuta* means "pointed duck." We call it pintail. Owing to the eastern chipmunk's markings and its habit of hoarding food, it's named *Tamias striatus*, "striped storer." The tiny, nocturnal little brown bat is *Myotis lucifugus*, "the mouse-eared animal that flees from light."

Unfortunately, a lot of names are cobbled together from elements of both Latin and Greek, so they're difficult to translate with standard dictionaries. The best reference for this sort of thing is *Jaeger's Dictionary of Biological Terms*. It's in print but hard to find and expensive, so I'll translate the more interesting names as we go along.

The same is true of my occasional references to geologic time frames. Just about every dictionary has a geologic chart, but if you don't care to look them up or memorize just how long ago the Mesozoic was, don't fret; I'll supply the numbers to go with the names.

And finally, you'll notice that I frequently talk about range and home range in describing where and how animals live. The two terms mean essentially the same thing, except one applies to species and the other to individual animals. Range is the geographical area where a species can be found; it may take in entire continents. Home range, on the other hand, is the living space that individuals occupy. Depending upon the animal, a home range may be as small as an acre or two or as large as many square miles.

# I

# Strength and Stealth

# 1

# Brother Bear

An ancient Cherokee legend has it that bears descended from the people of a Cherokee village who chose to leave the toil and fret of human life and to live instead as foragers of the woods. Long black hair soon grew over their bodies. "Call us when you are hungry," one of them said to those who remained behind, "and we will come and give you our flesh. Do not be afraid to kill us, for we shall live always."

To the Pueblos, bears were particularly favored by the two principal gods—the Mother of Game and the Spirit of the West—and the Pueblo hunter who killed a bear became worthy of the company of warriors.

Many North American Indians believed that bears possess supernatural powers, especially powers of healing. Indians of most cultures believed, like the Cherokee, in some special kinship between bears and man.

The animals themselves best demonstrate where such notions came from. The skinned carcass of a bear looks remarkably like the body of a man, and some of a bear's bones have an eerily human look about them. Bears often stand upright and sometimes shuffle along for short distances on two legs like well-fed burghers in shaggy overcoats, homeward bound from a fine dinner where the wine flowed freely.

And bears, like man, are omnivorous, finding a comfortable livelihood on virtually anything that walks, crawls, swims, flies, or grows.

Science validates something of what the Indians recognized as humanlike in a bear's posture and gait. We're both plantigrade animals, walking with our five toes and the soles of our feet flat on the ground. Dogs and cats, on the other hand, are digitigrade, which is to say they walk on their toes.

Even so, bears and dogs and all the other members of the order Carnivora share a common ancestry, one that despite our similarities and the romance of legend, does not include man. The fossil record suggests that all of the carnivores descended from ancestral mammals that lived during the Paleocene, 60 million years ago. As various forms developed and diverged, the group that includes bears, raccoons, dogs, and sea lions branched off from the others. More direct ancestors of the family Ursidae, the bears, branched off again about 30 million years ago, during the Oligocene.

Bears existed on the European continent at least 15 million years ago. The earliest evidence of their presence in Asia dates back about 4 million years. They appeared in North America about a million years later and in South America about a million years after that. Africa is something of a puzzle. Brown bears have lived in the Atlas Mountains of North Africa since the Pleistocene, about 2 million years ago, but the only other record of bears on the entire continent comes from a 5-million-year-old site in South Africa. How and when the bear

*Agriotherium* got to South Africa and why it failed to survive is anyone's guess.

As the earth has undergone its continual geologic and atmospheric changes, many species of animals have appeared, thrived for a while, and then gone extinct, including some species of bears—like *Ursus spelaeus*, the huge plant-eating cave bear of the Pleistocene ice ages. Nonetheless, bears still inhabit all of the world's continents except Antarctica and Australia.

At present, taxonomists recognize eight main species of the family Ursidae worldwide, organized in three subfamilies.

The subfamily Ursinae is the largest, comprising six species: the Asiatic black bear, which lives all across southern

Asia from the Middle East to Japan and Taiwan; the Malayan sun bear of southeast Asia and the islands of Indonesia; the sloth bear of India, Nepal, Bangladesh, and Sri Lanka; the brown bear of Europe, Asia, North Africa, and North America; the polar bear of the high Arctic; and the American black bear of North America.

The other subfamilies include one living species apiece—the spectacled bear of South America and the Asian giant panda. There's some dispute among mammalogists whether the panda is a bear at all or whether it should be classified with its other close relative, the raccoon.

Bears are adaptable animals, generalists in terms of where they live and what they eat. The three species that occur in North America occupy virtually every sort of habitat the continent has to offer, from southern hardwood forests to boreal conifer woods, from high mountain meadows to bleak tundra to the vast polar icepack. Nevertheless, bearlike nature is largely the same among all species everywhere, marked only by such differences in behavior and physical equipment that particular habitats demand.

Bears have large heads, massive bodies, short, immensely powerful legs, stubby tails, and small eyes and ears. Their basic tool kits comprise ten thick, recurved claws, a formidable set of forty-two teeth, and keen intelligence. The claws are capable of heavy-duty tearing and digging, and yet bears are dexterous enough to delicately hook a morsel of food from a tiny crevice with a single claw. The teeth serve a similarly wide range of functions: there are incisors and long canines for gripping and tearing, and broad, flat molars for grinding plant material. Owing to its varied diet, a bear has no carnassials, the flesh-shearing cheek teeth that are highly developed in other, more specialized carnivores.

Lots of animals have sharper vision and better hearing; not many have a more sophisticated sense of smell. It's said they can smell the contents of an unopened tin can. Certainly, the brown bears I've been around in Alaska have had no difficulty catching my scent, even at several hundred yards when the wind was right.

They lumber along, looking bulky and slow, but bears are remarkably agile when they need to be and can readily outrun a man. A polar bear can run down a caribou in a short sprint.

The big bears are the largest living carnivores on earth, though their size tends to be exaggerated in the popular view. Adventure-magazine pictures of bears towering fifteen feet over some doughty woodsman are only an illustrator's fancy, but even if they don't grow to the size of locomotives, bears are big enough.

The black bear is the smallest of the North American species, standing perhaps a yard tall at the shoulder and measuring sixty to seventy inches from nose to rump, with another four or five inches of tail. They weigh 200 to 300 pounds on average, though some have been reported at more than 500 pounds.

Polar bears are considerably larger than blacks, averaging six to eight feet long and five feet at the shoulder. The average male weighs about 1,100 pounds. Males of all species, incidentally, are larger than females.

Brown bears show the widest variations in size, depending upon where you find them. They can range from five to nine feet nose to rump and from three to five feet at the shoulder. In southern Europe, they average about 160 pounds; 400 to 500 pounds in North America, Siberia, and northern Europe. The real heavyweights, the largest bears on earth, live on Alaska's Kodiak and Admiralty islands. There they may grow as large as 1,700 pounds and are awesome animals, indeed.

Black and brown bears show such wide ranges of color and size that taxonomists have fussed for years over how they should be classified, one school of thought arguing that they represent several distinct species, another that the variations are simply different color phases of the same animal, worthy of subspecies status perhaps but still basically the same. Black bears in the East and South generally have glossy black pelts; in the Rockies and other parts of the West, they're often a reddish cinnamon-brown. In Canada and Alaska, they might be black, chocolate, blue, or even ivory-white.

Brown bears have been the subject of even more taxonomic wrangling. The dispute generally has centered more on size than color, but color, too, has been the bone for some contention. Brown bears usually are dark brown, but they may also be cream-colored or nearly black. Those in the Rocky Mountains commonly show a good deal of silvery white in the guard hairs of their shoulders and backs, hence the common names *grizzly* and *silvertip*. They once were thought to belong to a species separate from the other brown bears, just as the behemoths of Kodiak Island were considered to be substantially different from their cousins of more modest size.

At present, the lumpers among the scientific community have the upper hand on the splitters, and all brown bears—grizzlies, Kodiaks, and the rest—are classified simply as *Ursus arctos*. Similarly, black bears are *Ursus americanus*, regardless of what color they happen to be.

Unlike the others, specimens of *Ursus maritimus*, the polar bear, are all about the same size and color, essentially white—bright and pure after the annual molt, more yellowish in summer.

Except for mating pairs and females with cubs, bears are solitary animals, although they don't seem to mind running into one another now and then when home ranges overlap. Flying over the mouth of the Battle River in Alaska a couple of years ago, I counted eleven brown bears in what couldn't have been more than about four square miles.

◆

Brown and polar bears sometimes gather in large numbers around some particularly good source of food, whether it's a garbage dump or a riverful of salmon during the spawning run. In the breeding season, though—which comes in late spring and summer—males of all species will fight over females.

Bears reach sexual maturity at four to six years of age. Brown and polar bear females come into estrus every three years, black bears every other year. The mating period generally lasts about three weeks, and then the erstwhile lovers go their separate ways. Once in the female's uterus, the embryos develop in a relatively short time, ten or twelve weeks, but she may carry her fertilized eggs in a state of suspended development for six to nine months. The process, which is called delayed implantation and which occurs in a number of mammals, seems designed to promote a high rate of reproductive success by timing births to the optimum moment.

For most bears, that moment is the depth of winter. They spend the autumn feeding heavily, building up a thick layer of fat. Between eating jags, they find some sheltered spot and prepare a den. Caves, burrows under stones or tree roots, hollow logs, or snowbanks all seem to serve equally well. At the onset of cold weather, each bear takes to its den and sleeps the winter through.

Mammalogists disagree over whether bears truly hibernate. The sleep is not a deep one by the standards of such world-class hibernators as woodchucks, ground squirrels, and the meadow jumping mouse—that, even in the mild climate where I live, conks out in September, stays out for as long as seven months, and wouldn't wake up if you played ping-pong with it. Bears sleep more lightly, are easily roused, and sometimes wake up on their own if the weather turns warm. Their bodily functions continue working during the winter, and their temperature doesn't drop substantially below normal. Those who argue that bears truly do hibernate cite the heart rate, which drops to less than half its normal pace.

◆

In the warmer latitudes, many brown and black bears don't go into the winter sleep. Neither do polar bears, as a rule, although pregnant females do, because North American bears give birth in their sleep.

Depending upon where they live, the birthing will happen sometime from November to February. She may have only one or as many as four cubs, but it's all the same to a female bear. She is a devoted parent, and the bond between mother and young is the strongest and virtually the only social bond that bears form. Which is well, because newborn cubs need all the care they can get.

They are born blind and naked, although polar bear cubs come dressed in a bit of fur. All of them are born tiny as well. Black bear cubs are the smallest, averaging about nine ounces at birth. Brown bears, the largest of all as adults, weigh about a pound, and polar bears, born into some of the harshest habitat on earth, check in at about twenty-one ounces.

How long the family group remains together depends in some measure upon their species. Black bear cubs remain with their mothers through their second winter. Young polar bears generally strike off on their own when they're about two years old, brown bears when they're three or four.

Regardless, females with cubs are fiercely protective and extremely dangerous to anything they might perceive as a threat to their young. I suspect that's what has motivated the majority of bears that have mauled or killed people.

Which is not to say that such attacks are either frequent or inevitable. Unless closely confronted, or their cubs molested, sows seem more inclined to avoid people. On the same day I saw so many bears around the Battle River, two friends and I were fishing when a female and three nearly grown cubs showed up, working their way downstream. When she became aware of us, she hustled all three out of the water and led them around us in a wide berth—which was fine with us.

Among the things that young bears learn during their time with mother is what's good to eat and where to find it. For black and brown bears, what's good can be almost anything. At least three-quarters of a black bear's diet is vegetable—roots, grass, nuts, berries, acorns, fruits, and the inner bark of trees. They also eat a fair share of ants, bees, crickets, grasshoppers, fish, frogs, bird eggs, small rodents, the occasional fawn, and any kind of carrion they happen to find, including other bears.

Brown bears, too, are mainly vegetarians, fond of grass, sedges, roots, tubers, moss, berries, almost anything that grows. But they're carnivores as well and will eat any animal from ants, mice, marmots, and fish, to moose, elk, caribou, deer, mountain sheep, goats, and black bears—which is one good reason why there never have been a lot of black bears in places where browns are plentiful.

Polar bears are more specifically carnivorous. Vegetation is hard to come by on the pack ice where they spend most of their time, so polar bears don't eat much plant material—although berries are an important food source for those that do wander inland during summer and fall. Though their primary food is the ringed seal, ambushed or stalked on the ice, the bears also eat sea birds, fish, the carcasses of marine mammals stranded on the ice, caribou, and various small land mammals.

Other than size, physical differences among North American bears are slight and reflect adaptations to their environment. Polar bears are top-notch swimmers; their bodies therefore are more streamlined, their necks longer, heads smaller and more flattened, their forefeet huge and oarlike. The soles of all four feet are thickly furred for traction and insulation.

Black bears are forest dwellers and expert tree-climbers. They, too, swim well, though their feet are proportionately smaller than the others'.

Brown bears are not strong swimmers, and their long front claws aren't well adapted to tree climbing. Neither lack is any disadvantage. They have little reason to swim, and when you're the size of a brown bear, you don't worry about climbing trees.

Unlike the Indians, who never existed in large enough numbers to offer much competition for living space, white men have always had an uneasy time with bears. When the first Europeans arrived, black bears ranged over much of the continent, from the central plateau of Mexico to northern Canada, living wherever there was forest. Brown bears were abundant from the Great Plains westward, from the mountains of Mexico to the subarctic tundra. In the North, their range extended east of Hudson Bay.

Human presence now has placed stricter limits on the bear's horizon. Even the polar bear has suffered from the encroachments of man. Brown bears are gone from the great central grasslands and, except for small populations in the West, are nearly gone from the entire lower forty-eight states. Strong populations still survive in Alaska and Canada. Black bears, too, are gone from many of their former homes but have extended their range in some places as brown bears have disappeared. Several states have programs under way to restore black bears in suitable habitat, and some of those programs have been successful enough to achieve not only healthy but also huntable populations.

Bears are big animals, and they can be dangerous. Every year, some encounter between bear and man comes off badly, for reasons that usually range from ill luck to ignorance to patent stupidity. How well we coexist in the future will largely depend upon how much moral fiber we demonstrate in leaving some room for the bears and giving them enough slack to be bears in it.

No bears have lived in my small part of the world for a hundred years or more—though I'm told a black bear has been seen in this county, and I'm inclined to think it's probably true.

◆

I sometimes come across them in the forests of southern Missouri, northern Arkansas, or northern Minnesota, when I'm hunting grouse or just poking around the woods. Not the animals themselves, as a rule, because black bears are shy creatures and want nothing to do with me. But I find their tracks and their scat, some old, some fresh, and it always sends a prickly feeling up my back, a little shiver that's partly atavistic fear and partly a feeling of delight to know that brother bear still shambles through these woods, still with us after all these years.

◆

# 2

# Sympathy for the Devil

In November 1573 Gilles Garnier came before the Court of Parliament of Dôle, in the old French province of Franche Compté. A reclusive, gray-bearded man of middle age known locally as the Hermit of St. Bonnot, Garnier stood accused of attacking several children in the countryside around Dôle, of having killed at least three, and of having partially eaten two of them.

Some peasants from the village of Chastenoy testified that on November 8 they had rescued a little girl from attack by a huge wolf and further swore that the animal's face had looked remarkably like the face of Gilles Garnier. Others testified to having found Garnier himself in a thicket near the village of Perrouze on the Friday before the feast of St. Bartholomew, crouched over the body of a boy, apparently prepared to begin a feast of his own.

Garnier confessed to these crimes and others, admitting that at certain times he roamed the country in the form of a wolf.

The court had on September 13 authorized the peasants of all the villages around Dôle to "assemble with pikes, halberts, arquebuses, and sticks, to chase and to pursue...to tie and to kill, without incurring any pains or penalties" any and all werewolves infesting the area. The justices were particularly irritated with Gilles Garnier for having performed some of his cannibalistic deeds on meatless Fridays and summarily ordered that he be dragged to the place of public execution and there burned at the stake.

At the time scarcely anyone in Europe would have harbored the slightest doubt that the testimony—including Garnier's transformation—was anything but literal truth nor that the sentence was anything but just. In fact people all around the Northern Hemisphere, from the Japanese to the Navajo, from the Arabian peninsula to the Canadian Arctic, would have considered the case of Gilles Garnier as unremarkable as grass—if for no other reason than their ancient conviction that any monstrous or bloodthirsty act was well in character for the wolf.

It's an old fear indeed. By the time *Australopithecus*, the earliest certain hominid, appeared some 2 million years ago, the wolf already roamed the northern half of the world, descendant of 120 million years' evolution from the primitive, flesh-eating creodonts of the Cretaceous. It evolved through the Miacidae, the ancestral family form from which all modern carnivores derive; through such later, more doglike ancestors as *Tomarctus*, parent of both the wolf and the fox; and finally, 15 million years ago, began developing on its own to become the most widespread land-living mammal, apart from man himself, to ever inhabit the earth.

According to considerable fossil evidence animals of the dog family originated in North America and from here dispersed around the globe. When *Homo erectus*, the earliest man of our genus, migrated out of Africa about a half-million years ago he found wolves in abundance throughout Eurasia. How they all got along we'll never really know. At the least, wolves competed with the new, two-legged predators and possibly even dined on a few. One thing, however, seems certain: Justified or not, early man lived in fear of the wolf, and proof of that lies in the atavistic chill that skitters up the back of even space-age man when the night echoes the distant howl of a wolf.

Filtered through the gauzy scrim of several hundred thousand years of imagination, that eerie, trenchant sound speaks of danger and dread, of a wildness so incorrigible and absolute as to reach the border of evil itself. As the author of a twelfth-century Latin bestiary puts it, "The devil bears the similitude of a wolf: he who is always looking over the human race with his evil eye, and darkly prowling round the sheepfolds of the faithful so that he may afflict and ruin their souls."

Small wonder that few have chosen to look behind the dark curtain of reputation to ask what manner of beast might truly lie beyond. To those who have, the view is at once familiar and surprising.

Wolves are the largest wild members of the Canidae, the dog family, although actual size varies according to where you find them. The farther south they live the smaller they are. Wolves of the Arabian peninsula measure scarcely a yard from nose to rump, and Mexican wolves are about the same size. The largest of them live in Alaska and western Canada, where males might weigh a hundred pounds or more, stand three feet tall at the shoulder, and measure six feet from nose to tail-tip. Females typically are a bit smaller. The largest wolf on record weighed 175 pounds.

They come dressed in a double-layered coat, a dense growth of fine underfur beneath a somewhat thinner stand of longer, stiffer guard hairs. The outer coat gives the animal its color, and coloration in the wolf covers a wider range than virtually any other mammal—from almost pure white to solid black and nearly every shade of cream, buff, tawny red, brown, and gray in between. Gray seems to be the most common color, although it's actually a mixture of black, white, brown, and gray hairs combined with buffy legs, underbelly, muzzle, and ears.

Such color variation and the fact that wolves originally occupied every sort of habitat in the Northern Hemisphere except desert and tropical forest accounts for their multitude of names—gray wolf, white wolf, timber wolf, prairie wolf, arctic wolf, and others. That coyotes often are called "brush wolves" only thickens the confusion. Current consensus is that a wolf is a wolf, *Canis lupus*, no matter where it lives or what color it happens to be, and that a coyote, while an extremely close relative, belongs to a fully separate species, *Canis latrans*.

Still, lest things get too simple, taxonomists continue to argue over subspecies. The exact number depends upon whose assertions you care to accept, but most agree on two dozen subspecies in North America and eight more in Eurasia. Some of these denote populations now extinct, and virtually all are based on differences in coloration and skeletal size.

And there's the matter of *Canis rufus*, the red wolf, a coyote-sized, reddish-coated animal formerly found all across the southern United States from Florida to the Southwest. Although mammalogists long recognized it as a separate species complete with three subspecies, both its origin and its fate are now unclear. Before about 1970 the majority of scientific attention focused on what appeared to be the animal's steady disappearance. The consensus was that persecution and ecological disruption by humans had allowed coyotes to invade the red wolf's range and to genetically swamp the species by interbreeding.

But even as the U.S. Fish & Wildlife Service and several state conservation agencies struggled to protect the remaining "pure" red wolf populations, two Harvard researchers applying sophisticated computerized statistical techniques concluded that the red wolf of Florida and the Mississippi valley simply is yet another race of *Canis lupus*, while the western subspecies is a hybrid of wolf and coyote parentage. Shortly after, another researcher from the Smithsonian Institution insisted that his study of the western subspecies showed the red wolf to be a race of coyote and not a true wolf at all.

◆

The most persuasive opinion at present is that all red wolves are fertile hybrids originally derived from interbreeding between wolves and coyotes. The question still isn't settled, and the red wolf, whatever it is, meanwhile continues to fade toward extinction.

The most striking differences between the wolf and its nearest kin have less to do with physical characteristics than with emotional and behavioral ones. Both coyotes and red wolves are largely solitary creatures for whom the mated pair represents the basic social unit. Not so the wolf. *Canis lupus* lives in highly organized groups held together by enormously powerful emotional bonds.

Ever since early man first recognized their existence, the wolf pack has been regarded as the ultimate predatory force in nature, and through the embroidery of fancy, popular lore has invested the pack with preternatural cunning. In that view one wolf is simply a threat, but a pack is virtually a super-organism, diabolically clever and irresistibly capable of anything from mischief to bloody mayhem.

In more realistic terms, a wolf pack is a group of individual animals who travel, hunt, feed, and rest together, sharing a general lifestyle in which the group exerts a greater influence than any single member. Folktales about packs of fifty or sixty or more are just that; according to Dr. David Mech, who knows more about wolves than virtually anyone alive, the largest reliably recorded pack numbered thirty-six, but even twenty together is an unusually large group.

No one knows all the factors that regulate the size of packs, but Dr. Mech suggests that the parameters include the minimum number of animals that can efficiently obtain food; the maximum number that can be supported by the amount of food available; a number that does not exceed each individual's capacity for social bonding; and similarly, a number that does not exceed each individual's tolerance for social competition. However strongly these and other factors may bear—particularly the survival rate among pups—the typical wolf pack contains from two to eight members.

If huge packs are the stuff of fairy tales, so also is the notion that they are nature's equivalent of street gangs. On the contrary, the pack is a family group comprising a mated pair and their one- or two-year-old offspring, and each animal is emotionally bonded to every other member.

This bonding is the cement that holds a wolf pack together. Ironically, the capacity for emotional attachment is precisely the trait we find so endearing in the dog—and *Canis familiaris* inherited it from its immediate ancestor, the wolf. The irony is that we persist in misunderstanding the wolf despite daily, intimate contact with an astonishingly wide range of wolflike behavior.

A dog's loyalty and devotion to its master is the crux of the man-dog relationship. This also is a key characteristic of the wolf. Packs are organized in a complicated social hierarchy headed by a leader, a "master," an authoritarian individual to whom all other members defer. The leader, usually a male, guides much of the pack's activity; he determines routes of travel, initiates hunting, takes charge when the pack is actively pursuing prey or fleeing from danger.

The leader's authority derives from his ability to dominate other pack members, so wolves' social system is based on the classic pecking order in which every individual occupies a certain rank. Among sexually mature wolves, separate social hierarchies exist for males and females. These operate within the overall pack structure, which lends further complexity to the animals' social relationships. Being a wolf among wolves is no simple matter.

So complicated a system could hardly function at all without high levels of communication, and wolves, like dogs, are wonderfully articulate. Although they vocalize in a variety of whimpers, growls, squeaks, grunts, moans, barks, and howls, the bulk of communication among canids in general is visual. Body postures and tail positions carry specific, sophisticated messages.

Only the primates have more expressive faces, and wolves communicate a remarkable repertoire of attitudes and emotions with their muzzles, lips, foreheads, ears, and eyes. (Which incidentally is why a hard stare is far more effective discipline for a dog than any amount of rump-swatting; all the Dr. Spocks of dogdom notwithstanding, the fabled newspaper method only sends a garbled message to the wrong end of the animal.)

Canine communication has well-developed tactile and olfactory components, and watching a group of wolves—or dogs, for that matter—perform all their various rituals is a humbling experience. Communication obviously is taking place, but the mechanics are so multifaceted and so subtle that one has to wonder if we humans aren't a bit too proud of our capacity for language. I make my living using the language and cherish my library full of books, but my dog doesn't seem the least bit handicapped by being unable to read or write a single word.

Its elaborate social structure serves to enhance the pack's survival and well-being and therefore that of each individual member. Wolves are strictly carnivorous and customarily prey upon animals considerably larger than themselves—deer, musk oxen, mountain sheep, caribou, elk, moose, and even bison—which naturally puts a premium on cooperative behavior. A big wolf is capable of single-handedly killing a moose, but even a deer is capable of killing a wolf, so as far as the pack is concerned there is both strength and safety in numbers—at least in numbers that allow the pack to function efficiently as a unit.

Popular myth supposes wolves to attack anything and everything that moves—to "rage about with tooth and claw," as the twelfth-century bestiarist I quoted earlier puts it. The truth is somewhat less melodramatic. Through most of the year wolf packs travel almost constantly, moving at a tireless trot of about five miles per hour, making the rounds of a home range that may be as small as seven or eight square miles during the summer or as large as five thousand square miles in winter. Since carnivores cannot simply stroll out to feed any time they feel like it, as plant-eating animals can, wolves are continually on the alert for prey; being opportunists of the first order, a pack is likely to show interest in any suitable prey animal it comes across, whether by chance, by scent, or by tracking.

If the encounter is close enough, wolves will test potential prey, but "testing" does not imply an attack nor does an attack imply a certain kill. From his observations of a wolf pack on Isle Royale, David Mech documented interaction between the wolves and 131 moose, which is the principal prey species on the island. All but 7 of the 131 escaped direct attack, either by detecting the wolves first, standing their ground, or simply outrunning the pack; of those 7, only 6 actually were killed.

Dr. Mech describes the wolf's typical hunting routine as a four-act drama involving a stalk, an encounter, a rush, and a chase. According to his observations, the curtain can come down at any time. Prey animals that get a substantial headstart on a pack or those that simply stand their ground are least likely of all to be attacked. A wolf can run at thirty-five or forty miles per hour in fifteen-foot bounds, but only in short sprints. They apparently are quick to judge when a chase is likely to prove fruitless, for they seldom pursue prey for more than a minute or so. Pitched battles similarly are not much to their liking, which has led to charges of cowardice from some and which seems to me a sign of keen intelligence.

When they do make a kill, wolves dine with hearty gusto. A wolf can pack away as much as twenty pounds of meat at a feeding, although the average is a far daintier five or six pounds. Nonetheless, they are not the wanton butchers we've made them out to be, a mistaken conclusion drawn by people who have found partially eaten wolf-kills without finding the wolves and therefore decided that wolves kill more than they actually eat. In fact they make efficient use of their food. A pack usually remains somewhere near a kill for several days, ultimately consuming everything except the skull, the largest bones, the hooves, and some of the hair.

◆

It doesn't fit the pronouncements of Mother Goose, the Brothers Grimm, nor the more macho outdoor magazines, but real wolves prey more upon very young, very old, and infirm animals than upon any others. The wolf's popular reputation for rapaciousness—a view once given credence even by some naturalists—simply makes no sense. No predator bent on wiping out its own livelihood could survive for long, and no clear-headed study has ever shown that wolf predation represents a significant limiting factor on any prey animal except in situations where man has played a fundamentally disruptive role in the first place.

Which is not to say the wolf has been without influence. For scores of thousands of years wolves were the most numerous and most widely distributed predators of large animals in the Northern Hemisphere, and their impact has been profound. Ironically though, most of the benefits of that influence have accrued to the prey. As wolves evolved into more capable killers, prey animals evolved better means of avoiding them. Natural selection is a thoroughgoing process, and organisms respond in a multitude of ways. Species of small prey animals—rabbits and mice, for instance—survive by being hyper-reproductive; larger ones become keen of sense and fleet of foot. Some live in herds, finding optimal security in numbers. Other responses are geographic or behavioral. Wild sheep and goats most likely live where they do mainly because their steep, rocky habitat is a good measure of protection against wolves. Defensive behavior among musk oxen, in which adults surround their calves in an outward-facing, rump-to-rump circle, probably evolved because of the wolf.

The point here is that while wolves are capable predators, they aren't particularly efficient. They fail at catching and killing prey far more often than they succeed, and if they choose the young or the weak it's because those are the animals they can kill with the least effort—not because wolves are cowardly or cruel or unsporting.

They prefer large prey, for obvious reasons, but they'll settle for less. Beavers, Dr. Mech says, are important prey in some areas, and wolves will hunt Arctic hares and even mice when nothing else is handy. If they must they can go several days without eating at all.

Another reason why big animals are a wolf's main target is that the pack sometimes has pups to feed. But like almost everything else in a wolf's life, even reproduction is fraught with complexities.

Wolves reach sexual maturity at the age of twenty-two months, but the biological imperative involves far more than just blissful dalliance. Courtship is a complicated affair that may last for days or even months. Their social structure is such that the dominant pair typically are the only ones to breed in any wolf pack, although the alpha male may at times choose not to and still retains his status. Subordinate members, however, are given a rough go of it, as the dominant animals usually disrupt their courtship activity, forcing them either into abstinence or departure—which probably is an important factor in regulating the size of packs and in establishing new ones.

When it is successfully carried out, mating takes place as early as January or as late as April, depending upon where the animals live, and gestation takes about sixty-three days. A month or so before giving birth, a pregnant female seeks out and begins preparing a den. It might be newly dug, an enlarged foxhole, an old beaver lodge, a hollow tree, a cave, or just a shallow scrape in some protected spot. Her only concern apparently is that it be of a proper size and have good drainage, for she prepares no bedding.

She takes to the den a day or so before the birthing and as each pup is born breaks the amniotic sac, severs the umbilicus, licks the newborn clean and dry, and nudges it toward the warmth of her belly. Newly whelped wolf pups weigh about a pound each, are covered with dark natal fur, and are deaf and blind. Six is an average litter.

During the first few days the female stays with her pups almost constantly while her pack mates provide her with food.

The pups' eyes open after about two weeks; their hearing begins to function about the twentieth day. They emerge from the den at three weeks.

Their most important business at this point is learning to be wolves in the company of wolves, finding their places in the social structure that will govern the remainder of their lives. They establish status among themselves by endless chasing, shoving, wrestling, and biting. Dog pups interact the same way and for the same reason, although less intensely owing to their zillions of generations' removal from the wild. Wolf pups may actually injure one another, but there is much at stake. Settling the matter of social roles can make the difference between life and death for a wolf, and pups usually have things sorted out by the time they're a month old.

At the same time the pups also establish relationships with the rest of the pack, learning their roles in the community and forming emotional attachments. Feeding seems to be a key mechanism in accomplishing both. All members of the pack help care

for and feed the pups once they're weaned, which occurs about two weeks after they leave the den. This clearly promotes emotional attachment among the pups and the older animals, but the method of feeding adds yet another element.

Wolves do not simply drag dead or incapacitated prey animals to their young as noncommunal predators usually do. Instead the adults gorge themselves on their kills, carry food back in their stomachs, and the pups have to beg to be fed. They lick and mouth the lips and muzzles of the adults, who respond by regurgitating whatever they've eaten, which the pups promptly eat in turn.

Not a pretty notion, but it serves an extremely important purpose. Begging is submissive behavior, and submissiveness is vital to the wolf's social structure. Having proper behavior instantly rewarded with food is powerful conditioning, and the pups absorb it well. The same sort of mouth-licking and nuzzling stays with them lifelong as a submissive gesture toward a more dominant wolf.

By the time the pups are eight or ten weeks old the pack is once more on the move, traveling among a series of rendezvous sites and spending several days at each one. These sites, often near bogs and swamps, apparently are places the pack considers secure nurseries for the pups while the adults go off to hunt. By October the pups are nearly full grown and begin traveling with the others.

For all the clannishness among pack members, wolves aren't nearly so fond of outsiders. Strange wolves are likely to be chased off or even killed, and packs seem to take pains to stay out of one another's way, marking the boundaries of their home territories with scent posts and advertising their presence by howling.

Nothing is so peculiarly lupine nor so quintessentially wild as the low, mournful, drawn-out wail that one old northern trapper described as "a dozen railroad whistles braided together." Human ears can hear it at a distance of five miles or more in still, cold air. Packs are fond of choral howling, which they perform with great excitement and tail wagging. They may sing in preparation for a hunt (although they remain silent during the hunt itself), as a means of reassembling scattered pack members after a chase, to help maintain acceptable distance between themselves and a neighboring pack, or simply because they can.

In any event, few sounds in nature are so deeply moving nor so fraught with preconception. In the old days no horror movie was complete without a wolf howl or two in the sound track. Even if the animals themselves never appeared, their voices alone were enough to evoke the proper atmosphere of foreboding and dread among viewers nurtured on the image of the Big Bad Wolf.

Human mythmaking is replete with such images, reaching hundreds of generations into the past. Perhaps because we recognize something admirable in the wolf its reputation is not wholly black. In Christian legend a wolf guarded the severed head of the martyred St. Edmund. St. Francis of Assisi induced a wolf to live amicably with the villagers of Gubbio, and two Irish saints, Maedoc and Herve, lived on friendly, cooperative terms with wolves. Other myths tell of wolves rearing human children; Romulus and Remus, legendary founders of Rome, probably are the most famous.

But for every benevolent or even indifferent wolf in myth and legend a dozen more play the villain—from false prophets to witches to evil spirits to Satan himself. Such fear of the wolf is a curious thing, but more curious still are the people who, in their own minds at least, become wolves.

The folklore of virtually every human culture owns its share of were-animals, humans able through some magic to transform themselves into bears, lions, hyenas, crocodiles, leopards, tigers, jaguars, snakes, foxes, even birds. As most do so for malevolent purposes, such myth may be an attempt to explain or at least recognize the bestial side of human nature—the same duality central to both the medieval concept of the Great Chain of Being and to Freudian psychology.

Given man's terror of the wolf it's hardly surprising that the werewolf should stand out from the rest. To us it's the stuff of scary tales; to our forebears it was literally real. You can find serious discussions of werewolfism in the works of Plato, Herodotus, Pliny, Ovid, and Virgil—and it was an ancient idea even to them.

But the golden age of werewolfism came during the Renaissance, an era preoccupied with witchcraft and demonology. The witch hunters believed some werewolves to be human witches capable of transforming themselves; others were thought to be demons who invaded the animals' bodies. All were seen as a menace to the souls of Christian folk. With examples like Gilles Garnier to serve as cases in point, such arguments were persuasive enough to become self-fulfilling, and werewolf epidemics erupted all across Europe in the sixteenth century. During one outbreak a French judge reportedly condemned 600 werewolves to death.

Dozens of "documented" cases like Garnier's survive. In one bizarre episode in 1541 a self-confessed werewolf explained his thoroughly human appearance by assuring the judges that his wolf pelt grew inward. Curious to see for themselves, the magistrates ordered his arms and legs amputated and flayed. The subject unfortunately expired before they got back to him with a decision.

Modern psychiatry recognizes lycanthropy—a name derived from the Greek myth in which Zeus transforms Lycaon, King of Arcadia, into a wolf—as a delusional form of paranoia or schizophrenia and describes the great epidemics as mass hysteria, not unlike the St. Vitus's Dance, the *tarantella,* or other historical instances of collective disorder. A few clear heads came to similar conclusions even during the Renaissance. In *Discoverie of Witchcraft,* written in 1584, Reginald Scot suggested that insanity, not shape-shifting, explained the werewolf. Insanity itself, however, was at the time widely thought to originate from demonic possession, so superstition still prevailed.

◆

By the nineteenth century, enlightenment had progressed to the point where scholars generally agreed that Scot was right. The first serious study of the subject in English appeared in 1865, a delightful treatise titled *The Book of Were-Wolves, being an account of a terrible superstition*, by the Reverend Sabine Baring–Gould (who also composed and wrote "Onward, Christian Soldiers"). The good reverend describes the malady as a combination of hallucination and bloodlust and declares, "The cases in which bloodthirstiness and cannibalism are united with insanity are those which properly fall under the head of Lycanthropy." That he doesn't cite any cases where bloodthirstiness and cannibalism *aren't* considered insanity suggests that enlightenment still had a way to go.

So, unhappily, does our view of the wolf itself. Historically we have gone hell-bent on wiping it from the earth, and we have very nearly succeeded. Few if any wolves remained in England by the middle of the thirteenth century. Five hundred years later they were gone from Ireland, Scotland, most of Europe, and all but the remotest wilderness of Asia.

In the New World the story is much the same. When European man arrived wolves lived nearly everywhere on the continent. How many there were in presettlement America is impossible to tell, though the wolf's lifestyle and social structure generally doesn't promote dense populations.

But they must have been abundant nonetheless. In 1911 Ernest Thompson Seton's reported records of the Hudson's Bay Company show about 278,000 wolf pelts sold in the fur trade from 1821 through 1908. The greatest numbers of these, as many as 14,000 in a single year, were taken between 1843 and 1868, a period corresponding with the introduction and widespread use of strychnine. In those years the company bought a total of 230,750 wolf pelts, almost 8,900 per year on the average.

Considering that the numbers account only for wolves killed for the fur trade within Hudson's Bay Company's sphere of influence, which by the 1820s didn't include much of the

United States, these more than a quarter-million animals could have been only a small fraction of the North American population.

By the mid-nineteenth century, however, wolves had been killed off along the Atlantic coast. Except for small populations in northern Minnesota, northern Wisconsin, and the Upper Peninsula of Michigan, they were gone from the entire eastern half of the continent south of the St. Lawrence River by 1914, victims of poison, bounties, the steel trap, and the rifle. In 1915 the federal government launched a full-scale effort to destroy predators of all kinds, and in less than thirty years the wolf had all but disappeared from the West.

Even considering man's age-old fear of the wolf, one must wonder at such a vicious pogrom. Livestock predation has been a problem in some areas from time to time, mainly because domestic animals are far easier prey than wild ones. But as is usually the case, actual numbers of livestock lost to wolves bear little relation to the numbers claimed or to the numbers of wolves slaughtered under the banner of animal husbandry.

We fear the wolf less, I suspect, for the welfare of our grazing herds than simply for ourselves. If the Big Bad Wolf was set on eating the Three Little Pigs and Little Red Riding Hood, he surely must be after us as well. You don't have to look far to find all the horror stories and trapper's tales you care to hear about nighttime travelers on lonely roads waylaid and eaten by ravening hordes of wolves. You'd think European wolves must be especially bloodthirsty judging from the number of stories that have them killing whole villages, or from the classic melodrama of a pack chasing a fleeing *troika* across the frozen waste while terrified riders toss out one family member after another to lighten the load and distract the slavering, fiery-eyed devils in pursuit.

They may be great fireside stories but virtually none is true. Researchers have gone to great pains to authenticate these events and have come up almost empty-handed. European wolves no doubt have attacked humans on occasion and perhaps have even killed a few, but the facts are nothing like the folklore.

◆

In North America the evidence shows even more strongly that the wolf has gone down on a bum rap. There never has been an authenticated instance of a wolf killing a human being anywhere on the continent, and there are only three documented cases of contact between a wolf and a human that even resemble attack. One was by an obviously rabid animal, another simply an aggressive leap. The third, reported in the best penny-dreadful tradition by one of the more popular outdoor magazines, happened a few years ago in northern Minnesota, where there's a population of about 1,200 wolves, the largest anywhere in the lower states.

The facts, as it turned out, were ludicrous: A wolf jumped onto a teenage deer hunter's back, growled once or twice and then took off as if its tail was on fire. And why? Well, it seems the little comedy took place just at dark, when only owls and bobcats can see very well. The boy was bent over fastening a snowshoe or something so as to present nothing of an upright human profile. And he was drenched in bottled deer-scent, which is meant to mask human odor and apparently does. The only sensible explanation is that a wolf mistook young Nimrod for a deer, jumped on him and, when the lad screeched like a scalded cat, did what any healthy wolf does in the presence of man.

End of story, which no doubt garnered a fair number of bucks for the writer and the magazine and probably cost a few wolves their lives while the whole thing simmered under the guise of news. Researchers certainly do not agree on every question of lupine biology, but they are all of a mind on one point at least: Wolves are extremely shy of man and do not attack people—even when they would be fully justified in doing so. Biologists have crawled into wolf dens and kidnapped pups while the parents stood by, fidgeting and whining; have handled, ear-tagged, and even examined the teeth of snared wolves with scarcely a struggle; have lived for months in areas harboring more wolves per square mile than anywhere in North America and caught no more than a few fleeting glimpses of them. And these are men and women whose profession it is to live with the animals they study.

Where, then, lives the Big Bad Wolf? In our imaginations and nowhere else. The real wolf, to what may be our everlasting shame, lives on the ragged edge of survival, a magnificent animal brought almost to extinction by fairy tales in the dark. Lest we have a care it might one day soon live nowhere at all.

◆

# 3

## Cat of One Color

In the early 1920s, during one of his sojourns in Mexico and the American Southwest, British writer D. H. Lawrence met two Mexican hunters who had trapped and killed a mountain lion. In a poem written shortly after, he describes

> A long, long slim cat, yellow like a lioness...
> Her round, bright face, bright as frost...
> And stripes in the brilliant frost of her face, sharp,
>   fine dark rays,
> Dark, keen, fine rays in the brilliant frost of her
> face.

Less than a generation before, Theodore Roosevelt had called the same creature "Lord of stealthy murder...with a heart both craven and cruel."

It's difficult to imagine more antithetical views of one animal, difficult to decide which is more nearly correct—the Englishman's sensual preoccupation with that frost-white face or the sturdy American's standard of manly courage.

What isn't difficult is understanding why such divergence should exist.

The mountain lion is peculiar to the New World, once ranging from British Columbia to Nova Scotia, through every state in the Lower Forty-eight, south through Mexico and across the isthmus, all the way to Patagonia. In the north, it is variously known as mountain lion, puma, panther, painter, cougar, and catamount; in Spanish-speaking regions it is simply *león*; in the scientific community, *Felis concolor*, "cat of one color."

The question, as drawn between Lawrence and Roosevelt, centers on exactly what shade that one color truly is—angelic pale or the shadow of menace. It remains unresolved even now.

The New World colonists brought with them the peculiarly European notion that everything must belong to someone, and also its corollary—whatever belongs to someone is entitled to every means of defense. No other single notion has been more influential in shaping the nature of life in America, and its effects continue unabated. (If you doubt this, tell me why the legal community continues to prosper on a burgeoning diet of litigation in which some form of property is the issue.) Thus our forebears viewed the great American wilderness according to whether it implied value or threat to the holy articles of ownership. What was useful was there for the taking; what remained incorrigible had to go, like the wolf and the bear and the buffalo and the lion.

Before the turn of the eighteenth century, Spanish priests in California offered a bull for each *león* the native peoples delivered dead to the missions. By 1760 Massachusetts Colony offered bounty on them. From 1820 to 1845 more than five hundred lions were killed in Centre County, Pennsylvania, alone—sixty-four of them by one man. The last cougar known in Pennsylvania was shot in 1891. In 1837 one Alphonso Wetmore complained that lions were still doing harm to the cattle industry in Missouri; the last one recorded in the state was killed in 1927.

Reviled as slaughterers of livestock and game, feared as a perennial threat to man, the great cats were methodically expunged from the East and the South, from the midwestern forest and prairie, and from the Great Plains until by about 1920 they scarcely existed at all except in a tiny enclave of swampland at the tip of Florida and in the deserts and rugged mountains of the West. Even there they suffered the hatred of stockmen and so-called "conservationists" looking for a scapegoat on which to blame dwindling populations of deer and elk. But for a sparse human population and an ability to occupy all-but-inaccessible habitats, the puma might well have gone the way of the wolf and disappeared from the United States almost altogether.

Through it all, any real understanding of its habits and life history remained as elusive as the cats themselves. Survey the literature and you'll find precious little on mountain lions published before the late 1960s. The most important work didn't appear in print until the '70s—particularly that of Maurice Hornocker and his colleagues. From these and other studies conducted since emerges an animal in some ways quite different from the minacious, wanton killer that has long prowled the American psyche.

There never has been any question that the mountain lion is the largest unspotted cat native to the Americas—a featherweight compared to tigers and African lions, but large enough. A big male might measure nine feet tip to tip (almost a yard of which will be tail), stand thirty inches at the shoulder, and weigh 150 pounds or more. Females are about 40 percent smaller on the average. Regardless of size or sex, cougars lack the sheer bulk of the other great cats and are essentially packages of sinew and muscle draped upon an astonishingly small framework. Disarticulated, an entire adult skeleton, skull included, will very nearly fit into a briefcase.

◆

Combining so much muscle with so little bone naturally creates a creature of prodigious strength, and the mountain lion is fully capable of preying upon animals many times larger than itself—whether a deer dispatched with one crushing bite or an elk brought down broken-necked with the wrench of a foreleg. Moreover, the cat typically carries or drags whatever it kills

to some sheltered, secluded place before feeding, a task that sometimes requires more strength than even the most athletic human possesses.

Even when the quarry is more modest—say, a hare or beaver, a porcupine, or even a mouse—the puma remains a predator of great finesse, master of stealth, quintessentially catlike in its methods. Because they do not possess the lung capacity for marathon chases, cats are stalkers and sprinters, infinitely patient, aided by superb hearing and sight and a wonderful ability to remain motionless for long periods. How close they prefer to get before launching an attack varies according to the size of the cat and the size of the prey; mountain lions are most successful against deer and elk when they can make their way to within about fifty feet.

Unless the victim is somehow cornered, the initial rush is often the lion's only chance, but it's usually enough, an uncoiling burst of astonishing energy and speed. All eighteen of a puma's toes are equipped with long, sharply curved, retractable claws. The hind claws provide purchase for the spring; those in the larger forefeet, augmented by a dewclaw on the inner side of each ankle, lend a tremendously powerful grip. So armed, the cat's first intention is to bowl the prey off its feet. If it succeeds, a kill is virtually assured.

If the prey is small or the lion is especially hungry, it will polish off the kill in a single meal. More often, though, a puma will first go for the delicacies—heart, liver, and lungs—satisfy its remaining appetite with perhaps a haunch, and cache the rest under a cover of leaves or snow or other debris. It then loafs and lazes somewhere nearby for a day or two before feeding again. Coyotes, bobcats, and ravens often filch a share in the meantime, and if the carcass spoils the lion will forsake it altogether.

Still, cougars aren't nearly so wasteful of their food as bobcats and seldom kill more than they eat. Under optimal conditions a healthy adult lion may take about thirty-five deer in a year's time, culling the old and infirm. But as with all large predators, survival depends entirely upon how many prey animals are available, how stable their populations are, and how mobile they happen to be. These and other factors combine to create a remarkably complex system of land tenure, which more than anything else dictates the lion's lifestyle and behavior.

Except for African lions none of the world's great cats are gregarious creatures, associating with others of their kind only in mating season or as family groups comprising lone females with young kits. Otherwise they live solitary lives in a certain area of land upon which they maintain well-advertised tenancy. These home ranges naturally vary in size and location according to the environment, the food supply, and in some cases the season.

In the western mountains pumas generally roam larger areas in summer. A female's summer range may cover eighty square miles, a male's more than a hundred. Both occupy winter territories of about fifty square miles. In all seasons, individuals maintain roughly the same distance from one another, as do those whose environment does not require seasonal alterations.

Which is not to say that each lion is absolute suzerain of its turf. Home ranges often overlap, females' more so than males'. Males' ranges may overlap those of several females, and more than one male may occupy virtually the same territory, although they still maintain their space by using different parts of the range at different times.

Successfully avoiding one another naturally takes some work, and pumas signal their presence with a system of scent- and sight-marking—depositing urine and feces, making scrapes, leaving glandular scents, and so on, freshening their signs frequently. North American cats seldom vocalize as a means of advertising residency, which is just as well for us, because they all have voices disagreeable enough to make water run uphill. A puma's scream sounds like a soprano having a hemorrhoidectomy with a garden rake. Mountain lions don't have the same hyoid apparatus as tigers and African lions, so they cannot roar. But they can purr.

If pumas expend a lot of fuss and energy keeping clear of their own kind, the rewards are worth the trouble. Establishing more or less exclusive claim upon the prey within a certain area enhances their survival rate as individuals. It also helps them avoid fighting, which is a good thing; among such powerful, well-armed and relatively soft-skinned animals, no party in an altercation is likely to get out undamaged. Curiously, though, fighting is common in some lion populations, almost nonexistent in others—a phenomenon that seems to be related to the relative abundance of prey and to the animals' ability to maintain stable territories.

Being well-fed, secure, and healthy in turn makes successful breeding more likely, and the fact that males' and females' territories often overlap helps foster a vigorous genetic mix. In the face of their punctilious solitude, one might wonder how big lions ever manage to make little ones; in fact, they get together by the same means they use to stay apart. At the onset of estrus, the females' urine becomes laced with pheromones, so the scent marks she leaves to identify her territory also let passing males know that she's not averse to a bit of scratch and tickle. She'll be at least two years old when her first estrus comes on, more often three, and therefore likely to be well established in a home range that will provide food for her young. Females can come into estrus at any time of year, but since the birthrate peaks in July, after ninety to ninety-six days of gestation, most breeding clearly takes place in late winter and early spring.

As males remain with females only briefly during the mating period, mother and kittens make up the mountain lion's only social structure. She may bear only one or as many as half a dozen; two or three is the average. The kittens are about a foot long at birth and weigh roughly a pound apiece, blind, dressed in buffy coats spotted with black, and sporting dark rings around their little tails. Their eyes open in about ten days and like most young cats' are bright blue. This soon turns to amber, and the spots and rings of their coats eventually fade into the subtle, tawny shades adults wear.

If the local food supply is adequate, the young ones grow quickly, first on milk and then, by six or seven weeks, on kills the female brings to the den. By then they will be about thirty inches long and weigh ten pounds. Soon they begin to accompany the female as she hunts. By the end of their first year they will be nearly full grown.

They may remain with their mother just over a year or for as long as two years. Population densities of both lions and prey animals are important factors in determining how long the group stays together. This early social life is schooling for the young, the period when they learn the crucial skills of hunting, and what they hunt may also affect the duration of their education. Northern lions appear to remain in family groups somewhat longer than those in the Southwest; elk, larger and more difficult to kill than deer, are more common prey in the north, so a longer tutelage clearly is beneficial.

Nonetheless, their days together eventually come to an end, certainly by the time the female comes into estrus again, if not before. The young may hang around with one another for a month or two after leaving their mother, but eventually they, too, break up to find territories of their own. They likely will travel twenty-five miles or more from their mother's range, but their success in finding homes at any distance depends as always upon both prey and the number of other lions in the region. Paradoxically for such solitary animals, pumas are very slow to occupy new habitat, even when it offers everything they need.

Their extreme territoriality and slow population growth, proven by Maurice Hornocker's landmark field studies in the 1960s and '70s, debunks the old myth of a country overrun by lions—just as the same studies clearly show that lion predation does not greatly affect big-game populations. Even so, we haven't quite figured out what to do with *Felis concolor*.

By the middle of this century, nearly every state where mountain lions live had elevated its status from vermin to game and thereby provided effective means of controlling their numbers while at the same time maintaining healthy populations. The notable exceptions are Texas, where lions still are unprotected by any sort of regulation, and California, where official attitudes toward the puma have swung from one extreme to another.

California continued to pay bounty on lions until 1963, did not declare them game animals until 1970, and in 1972 placed them under full protection as a diminishing species. Over the next fifteen years all indications showed that the population was on the rise—suggested by significant increases in predation against livestock and suburban pets, in confrontations with humans, even by lions killed on highways, which seldom happens in any other western state. By 1987 professional conservationists recognized the need to reinstate a hunting season, but the move was thwarted by preservationist groups who convinced the state superior court that more research on the animal's actual status was needed. This political and legal obfuscation reached a peak in 1990, when California voters approved a ballot measure that banned sport hunting for lions altogether.

To my thinking the passage of Proposition 117 is a classic example of how an electorate largely ignorant of biological facts can be manipulated by appeals to emotion. The legislation itself, like other noble experiments, seems surely destined to fail. The problem in California, as elsewhere, isn't hunting. The problem is too many people converting ever more wildlife habitat into suburbs and freeways and airports. As the wilderness disappears, so does the mountain lion's living space and source of livelihood. Predators are opportunists, and lions are no exception; faced with dwindling numbers of deer, they're content to dine on sheep, cattle, and neighborhood dogs, and to live as best they can in what scraps of habitat remain available. Inevitably this brings them in conflict with human interests—which is precisely the problem that has always existed between man and mountain lion.

The one immutable law of wildlife is that species remain healthy only so long as their numbers do not exceed their habitat's capacity to provide a living. This is as true of mountain lions as it is of any animal, including ourselves. In the lion's case its predatory nature only compounds the problem. An overpopulated deer herd will quietly starve to death; predators will instead catch whatever is available, in the process violating the sanctity of private property, which in turn demands they be destroyed.

In short, an infinite number of mountain lions and an infinite number of people cannot occupy the same space, and since we seem to have little will for limiting our own numbers, we'll have to limit theirs. Complete protection of lions in areas heavily populated by humans is a fairy tale that no amount of legislation or emotional high-mindedness can make come true.

Actually Proposition 117 does not offer complete protection, since it proscribes only sport hunting. Lions responsible for livestock depredation may legally be killed and so presumably may those who develop a taste for French poodles and other items vital to the common good. Moreover, the bill mandates that California spend $30 million per year over the next thirty years in protecting habitat for lions and other wildlife.

All of this may be well-intended, but it is neither realistic nor practical. As the human population continues to grow, the number of lions that somehow run afoul of someone is certain to increase, and the State of California will be solely responsible for removing them. The expense of doing so, coupled with the additional $30 million earmarked for habitat, is likely to get short shrift in the face of the state's well-publicized budgetary dilemma. The sad part is that regulated sport-hunting is not only the oldest and most reliable wildlife-management tool but also an important source of conservation revenue.

Sadder still, the chief loser in the whole fiasco will be the lion itself. Even though the puma tends to be extremely shy in the presence of man, confrontations between lions and people already are growing more frequent in California, just as they are in Colorado and other places where human encroachment is especially severe, or where the local lions are unusually aggressive.

Lion attacks make good grist for the media, but on the whole you're more likely to be killed by a honeybee or a lightning bolt, to say nothing of being scragged by an automobile or your spouse. According to research conducted by a student at the University of California-Davis, only sixty-six instances of mountain lions attacking humans can be documented in the entire Western Hemisphere during the 236-year period from 1750 to 1986. Twenty-three of these proved fatal. Records naturally are incomplete, but the actual totals probably are not much higher.

Even so, cougars can be dangerous. Children are the most likely victims, simply because their size makes them appear more preylike. This is unfortunate not only for the handful of kids who may be attacked but also for the animal, since media reports of such incidents tend to foster the old myth of the mountain lion as murderous fiend. As incidents grow more frequent, public sentiment ultimately fastens upon the concept of prior restraint and demands that lions be extirpated from areas where people live or visit in numbers—indicting whole populations for crimes that few individuals *might* commit and forgetting that we are the invaders in their homes, not the other way around. Some level of protection obviously is crucial to the species' well-being; too much protection is apt to burden the animal with a public-relations problem it isn't equipped to handle. The best hope is to find some middle ground between hating the lion to extinction and loving it to death.

Objectivity is difficult when a mountain lion mauls or kills someone, but the fact is the animals involved cannot simply be labeled "bad" or "rogue," and killing them after the fact scarcely even treats the symptoms. Most attacks are perpetrated by young lions whose instinct for avoiding man is untempered by experience and whose instinct for finding a home-range territory is stymied by human presence.

Much the same habitat crisis bedevils *Felis concolor coryi*, a subspecies commonly known as the Florida panther. Once it ranged throughout the Southeast, from South Carolina to southern Missouri to Louisiana; now it is virtually unknown outside the Big Cypress Swamp of southern Florida, where the population is estimated at no more than fifty and possibly may be as small as thirty. Direct conflict with humans does not seem to be a factor, but that's about the only problem the Florida panther doesn't face. Only time will tell whether the captive-breeding and release program now in place can offset the small population's gradual genetic decline from inbreeding.

Elsewhere east of the Rockies the mountain lion is scarcely more than a maybe. Unconfirmed reports suggest that a few may now live in the Appalachians. Others are known to exist in Louisiana and Arkansas, and with luck the puma may once again roam the Missouri Ozarks. My old friend biologist-artist Charlie Schwartz believed a few lions already live in the state, and he and

I used to chaff one another over which of us would be the first to see one prowling a limestone ledge high above some south-Missouri river. I haven't yet, and now Charlie never will.

According to the best estimate anyone can make, 16,000 lions may still survive in the United States. Perhaps this is enough. Surely it is better than none. Although they are almost certainly doomed to dwindle in certain parts of the West, perhaps they will be allowed to return in the Midwest and the East, where renewed populations of white-tailed deer threaten to outstrip the habitat.

Oddly, the nearest I have come to seeing a lion here at home happened far from the southern mountains, in a federal wildlife refuge in northwest Missouri, just a stone's throw from the Missouri River. In the mid-1970s an amateur naturalist reported seeing a lion while birdwatching in the refuge. A lot of local "experts" scoffed, but I knew the man and considered him a reliable observer. That a lion might travel along the river-valley corridor seemed perfectly plausible. Naturally the prospect caught my fancy. As I lived nearby in those days and knew the refuge well, I spent days roaming the place, hoping for a glimpse and perhaps even a photo. I never saw the cat, but I found one perfectly formed footprint, broad as my hand, where it had crossed a dirt road and another partial track a few yards away at the edge of a cornfield. They were at least a day old, possibly two.

I have since been lucky enough to see lions, if only briefly, in the West, but the thrill of it was no greater than the feeling I got from those footprints, imagining how the sinuous, long-tailed animal must have looked padding silently through the night, ghostly beneath a sky wild with stars. I had no sense of being watched, as some people have described, but I wondered nonetheless if I weren't at that moment caught in the chilly amber gaze from a frost-white face. I hoped I was, pleased to know that some pulse of wildness, however faint, still beat in that tame farmland.

# 4

# Little Lions

And then there's the old story about the two lumberjacks in, let's say, northern Minnesota who spied a bobcat sitting in a tree. Doughty specimens of a breed not known for being shrinking violets, they decided to catch it for a pet, so one took off his mackinaw and started climbing.

In a minute or two he was out of sight among the spruce boughs, and presently a grand commotion erupted high overhead, a pandemonium of crashes, snarls, and half-coherent shouting. The jackpiner on the ground listened a while and then called up, "Sven, would you want for me to come up and help you catch him?"

"No, Ole," the other shouted down, "but if you don't mind I could use some help lettin' 'im go."

You'll hear a similar tale in the South; the punchline's a bit different, but the theme operates on the same contrast of a small, cuddly-looking package containing a ferocious personality that takes a decidedly dim view of being molested. And like most such stories, there's an element of truth in it. You'd be hard-pressed to find a handsomer animal in North America than a bobcat or its northern cousin, the lynx, nor one more willing to mind its own business—yet capable of roundly thrashing anyone foolish enough to invade its privacy at close quarters.

I don't know about you, but I find all those qualities highly admirable, and if I weren't so thoroughly canine in my own personality I wouldn't mind coming back someday as one of our little lions.

Both words describe them fairly. As it does with most predators, human imagination tends to make them larger than they really are. While the smallest bobcat is several times the size of a housecat, not even the largest lynx is a match in height and weight for the average Labrador retriever. It'd be more than a match in a bare-knuckle fight, but we'll get to the lion part later.

The bobcat, *Lynx rufus* (*Felis rufus* to some taxonomists), is a uniquely North American animal, living from southern Mexico to southern Canada. It measures from just under two feet to a bit over four feet in total length—four to eight inches of which is tail—and weighs from ten to forty pounds.

*Lynx lynx* (*Felis lynx*, if you prefer) is more widely spread around the Northern Hemisphere, with populations and subspecies in Eurasia and the upper regions of North America. The Eurasian lynxes are the largest of all, weighing sixty or seventy pounds, occasionally as much as eighty. Our version, *Lynx canadensis*, is roughly the same size as the bobcat but stockier. It also has a shorter tail, larger, furrier feet, long tufts of fur at the tips of its ears, and usually a more prominent set of muttonchop sideburns on its cheeks, especially in winter.

Their tails are the surest way of telling them apart. Not that you're very likely to get close enough to a live specimen of either one in the wild to spend much time examining its caudal appendage, but if you do you'll notice that a lynx's tail has a solid-black tip, top and bottom, while a bobcat's typically is marked

with some dark bars and bands, is white at the very tip, and all white underneath.

The surest way of *not* telling one from the other is to get tangled in the morass of common names they've accumulated. In parts of Canada the lynx is *loup-cervier* or in the Anglicized version, "lucivee." You'll hear the bobcat called lynx cat, red lynx, bay lynx, and tiger cat. Both are often referred to as wildcat or catamount. (This lovely old word was coined ages ago in Europe as a compression of "cat-of-the-mountain"—which itself may have come from the Spanish word *gatomonte,* "mountain cat." Most likely it originally denoted the European wildcat, a small feline similar in size and appearance to some domestic cats. In this country *catamount* is applied not only to the lynx and bobcat but to the mountain lion as well, which sometimes makes for epic confusion.)

To an extent, geography keeps things reasonably sorted out, for lynx and bobcat ranges do not greatly overlap, the exceptions being: along the Canadian border, to which bobcats have extended their range over the past seventy-odd years; in the Pacific Northwest; in parts of Wyoming and Colorado; and in northern New England. Otherwise the New World lynx is strictly a Canadian and Alaskan animal.

Their life histories and behavior are essentially similar while their differences are largely shaped by habitat. Not surprisingly, the lynx is specifically adapted to a cold climate—well insulated by a thick coat and equipped with big, fur-soled paws that serve as excellent snowshoes.

Both are almost exclusively carnivorous, but what they actually eat is very much influenced by where they live. The lynx is so devoted to preying on snowshoe hares that its numbers fluctuate along with theirs. Every ten years or so, having bred themselves to great abundance, hare populations crash; within a year or so lynx populations decline as well, mainly because a great many kittens die of malnutrition or outright starvation and because the breeding adults that do survive don't reproduce well.

As bobcat habitats are more diverse, ranging from deserts to mountains, swamps to deciduous forests, so is their diet. Small mammals make up the bulk of it—rabbits, mice, rats, squirrels, fawns, possums, and a quotient of domestic cats—but they also have a taste for birds, particularly wild turkeys, quail, and other ground dwellers.

Both the bobcat and lynx are capable of killing adult deer. Lynx probably take the most, especially in winter when the cat has an easy time moving over deep snow in which deer bog down. Bobcats, on the other hand, kill more domestic stock, simply because their range covers more farm and ranch land. Even so, neither one as a species significantly interferes with human interests, and a trapper who knows his stuff can with almost clinical precision remove the occasional individual that goes on a stock-killing spree. Cats, being no dummies when it comes to remembering the source of an easy meal, will keep coming back if they find ready access to a chicken coop, rabbit hutch, or sheepfold.

◆

And they tend to leave a clear signature. Years ago when my friend Willie Lyles was the conservation agent assigned to Cole County, Missouri, a man in Jefferson City called to report that "a pack of wolves" was making nightly raids on his rabbit hutches, killing them by the coopful, eating very little, and leaving the rest strewn about. It sounded like typical bobcat work, and Willie invited me to ride along when he went to have a look.

The cat might as well have left a neon sign spelling out his name. Everything in sight said "bobcat"—from the two-inch tracks in the dust showing the characteristically complex heel-pad print, to the screen-wire hutch tops slashed as cleanly as if with a straight razor. The old man thought it was a preposterous idea (as if a wolf pack living in the capital city of Missouri during the late 1970s wasn't), and he was astonished two days later when Willie showed him a twenty-pound male bobcat, snarling and hissing in the live-trap.

"Hell," the old boy said, "he don't look big enough to kill a rabbit. Looks more like a barn cat." But he kept a respectful distance nonetheless and didn't offer to help us carry the trap to the Jeep. Willie heard no more about dead rabbits, or wolf packs, after that.

The incident illustrates a couple of salient facts. Seen lounging in a zoo a bobcat does look like it would be equally at home sprawled on a haybale—or a sofa, for that matter—until you notice the chilly, indifferent stare in those sulphur-yellow eyes. Put one in a tight spot and you'll learn what those eyes mean. The sheer, taut power in that little body is hard to imagine unless you've seen it in action, but I'll tell you, I'd rather braid rattlesnakes than mess with a bobcat when his dander's up.

That there's no likelihood of it happening without deliberate intent on my part is the other important point. I don't know of a shyer, warier, harder-to-see animal. They live virtually, sometimes literally, in our back yards without our ever knowing it. They are shadows, wills-o'-the-wisp, mere flickers in the corners of our eyes, fleeting as an absent thought. Even in the hills where I live, which is woolly country, I can mention having seen one and from eight people out of ten get nothing more than a puzzled look and some comment like, "Bobcat? There's *bob*cats around here???"

Now this isn't to suggest that suburbia is home to unseen hordes of *Lynx rufus*, nor is the wilderness. Neither bobcat nor lynx is plentiful anywhere, partly because they aren't much fonder of their own kind than they are of us. Each individual occupies its own home range, keeping the boundaries carefully marked by various scent signs, and for the most part does not intrude upon the ranges of others.

The two species practice their isolationism a bit differently. The home ranges of female lynxes tend to overlap one another to some extent, and a male's range may overlap those of several females. Bobcats take the opposite approach: females keep their ranges strictly separate while the males don't seem to mind sharing parts of their turf. But even where ranges overlap, the resident cats, males and females alike, punctiliously avoid one another.

Their senses, particularly sight and hearing, are keen enough that chance meetings don't occur very often, and when they do the result is a discussion rather than a brawl, a confab carried on through body language—or more accurately, whisker language. With tails capable of little more than shorthand, bobcats and especially lynxes telegraph their moods through the motion and position of their whiskers, drawing them up or down, spreading them widely, or by some other gesture using them to say what's on their minds.

Their antisocial outlook naturally changes somewhat during mating season, which is a once-annually affair for both lynx and bobcat. And as with all cats, from domestics to African lions, the wailing, screeching, and caterwauling that accompanies courtship and copulation sounds more like combat than conjugation. The way bobcats carry on you'd think they were enjoying about all they can stand and then some. I've never heard lynxes, but I don't imagine they sound any better.

Still, as the saying goes, it works for them, and a successful lynx or bobcat mating produces a litter of two or three kittens eight to ten weeks later. They're born in a den under tree roots, in the hollow of a fallen tree, or in a cavelike crevice in outcropping rock. Young of both species arrive weighing about three-quarters of a pound, dressed in a suit of spotted fur, and with eyes tightly shut. Their mothers are their sole nurturers and caretakers.

Although she weans them after a couple of months the little family group will remain together until the next breeding season. The siblings may hang out with one another for a while longer, but eventually their instinct for solitude prompts them to find territories of their own. The quality of the habitat available and the number of other cats already established in it will determine whether they are successful or whether, for one reason or another, they perish in the quest.

If at that age they do die it will most likely be from an automobile, a trap, or a rifle. Coyotes, foxes, wolves, golden eagles, great horned owls, and male cats of their own kind kill a certain number of kittens, but not many predators are capable of taking down a full-grown bobcat or lynx. Distemper, rabies, and a variety of internal parasites afflict both species.

Bobcats are also a natural reservoir for cytauxzoonosis, which is highly fatal to domestic cats. My veterinarian friend Dr. Jim Wilsman tells me the disease is caused by a protozoan para-site thought to be transmitted by hard-bodied ticks. It attacks the spleen, liver, and lymph nodes, causing death from dehydration and internal hemorrhage four or five days after the symptoms appear.

Until 1976, when four cases were reported among housecats in Missouri and two in Texas, the ailment was known almost exclusively among African ruminants. But Dr. Alan Kocan of Oklahoma State University has demonstrated that cytauxzoonosis typically is not fatal to healthy bobcats—which suggests that the protozoan and *Lynx rufus* have been together for a very long time, long enough that bobcats have evolved an immunity to the disease.

That such discoveries are still possible in the age of information is hardly surprising; the sum of what we know about wildlife, or almost anything else, probably isn't a fraction of what remains to be learned—especially of such retiring animals as the lynx and bobcat.

As I said earlier, even a fleeting glimpse is a rare thing. I imagine tree-stand deer hunters see more bobcats than anyone else. You can sometimes attract them with a predator call if you set up in the right place at the right time, from sunset to sunrise. Otherwise getting a look is usually a matter of pure chance—like the most recent one I saw here on the farm.

I was driving my stepdaughter to school one morning, taking the back road to avoid the highway traffic. We were just up the valley, where the road begins to climb along the contour of the ridge, when a bobcat came rocketing up from the downhill side, crossed the road in two jumps, and leaped for the bank on the high side. The top of the cut is about seven feet above the road there, and he seemed to sail up in slow motion, hind legs cocked for his next bound. An instant later he hit the top of the bank and disappeared into the scrub oaks.

It happened so quickly that he was out of sight by the time I stomped the brakes and yelled something like "Omy*gawd*, did you see that?? That was in*cred*ible!! Did you *see* that???"

Beth bent forward and picked up the books that had slid off her lap, looked out at the empty road in front of us and then at me. "See what?"

"A bobcat!! My *Lord*, what a beautiful animal!!"

"I was looking the other way."

I continued to babble as we rolled on following the twisty road as it climbs to the ridgetop, muttering about how he probably was out later than he'd planned to be and was hightailing it for the rocky little gully to the east to hole up for the day. After a mile or so, Beth said, "Was there really a bobcat?"

"Oh, yeah. Big, handsome cat. He was gorgeous. Why?"

"Well," she said, "I have to tell you: I thought for a minute there you were losing it."

When you're fourteen I suppose adults who get overly enthusiastic about anything tend to seem a bit suspect, especially if it appears to be something that isn't there.

I assured her that I was in fact not losing it, but thinking about it later I had to admit to myself that sooner or later I probably will—and with any luck at all it'll be because of something just like that. I can think of worse reasons to go over the edge.

◆

# 5

# Red Rogue

*Old Mother Slipper-slapper jumped out of bed;*
*Out of the window she cocked her head.*
*She cried, "John, John, the gray goose is gone,*
*And the Fox is on the town-o.*

Traditional American Song

Somewhere in the mind of man, the fox is an ancient memory, a cunning spirit loose in the night. Through more than 3,000 years of myth and fable, the fox has been the clever rogue, a champion of wit, tameless as the wind. He is a marauder less malefic than the wolf and a reminder, half admired, of the wildness lying under civilization's thin veneer.

Foxes are older than civilization, older than man himself. Fossil evidence shows that canids—the family comprising foxes, wolves, dogs, coyotes, and jackals—lived in Europe and North America 40 million years ago or more. The foxes themselves are a diverse and widespread tribe composed of nine genera and twenty-two species; they occur naturally on every continent but Antarctica and Australia. Except for the tropics of southeast Asia and the southwestern Pacific islands, foxes occupy almost every kind of habitat from deserts to dry steppes, from savannahs to mountain plateaus, from suburbs to the polar icepack.

Six species occur in North America alone: the arctic fox, whose range is right around the Pole through Canada, Greenland, Iceland, and northern Eurasia; the swift fox of the Great Plains; the kit fox of the western coast; the gray fox, whose range covers most of the United States, Mexico, Central America, and south to Venezuela; another species of gray fox, *Urocyon littoralis*, that lives only on a half-dozen islands off the coast of California; and most widespread of all, the red fox.

To the majority of mankind in the Northern Hemisphere, the red fox is the archetypal fox, the familiar fox of folklore and legend, the animal from which our perceptions of what is foxlike derive. Even his scientific name, *Vulpes vulpes*, reflects this, for it translates from Latin simply as "fox fox."

He is the crafty Reynard of folktale and fable. He is the fox of Aesop, six centuries before Christ. He is the fox of Japanese folklore, possessed of such magic that he can take on the shape of *Koki-teno*, the beautiful fox-goddess. To the Algonquian-speaking Indians of Wisconsin, he is *wagosh*, namesake of the Fox Clan. He is Br'er Fox. He is the eponym of the fox terrier, fox trot, fox grape, foxhound, foxglove, foxhole, foxtail, and foxfire.

If *Vulpes vulpes* is troubled by all the sociological, etymological, mythic, and magical baggage we've put on him, he takes it with good grace. Besides, he is perfectly well equipped to live with us, in spite of us, or without us altogether.

He is a generalist in his requirements for living space, equally at home in the deep woods, on open grassland, arctic tundra, the patchwork biomes of farm country, suburbia, and even, according to studies done in England some years ago, in big cities. In fact the red fox may well be the most widely distributed terrestrial mammal on earth next to man, and he's amenable to transplant besides. Red foxes were introduced in Australia in 1868 and now range over most of the continent.

He isn't all that picky about what he eats, either. Animal matter makes up the bulk of his diet. Mice and rabbits are the staples, although he won't pass up a chance to take birds or other mammals if they're either small enough to kill handily or dead to begin with. He'll also eat insects and plants, and he seems to be particularly fond of persimmons, which to my mind is a clear sign of good taste.

Like all other canids from the wolf to the domestic dog, the red fox is an opportunist. Just as my highly bred Brittany isn't above snatching a steak from the kitchen counter or rooting through trash like the scruffiest mutt in mongreldom, a fox isn't about to pass up an easy meal if he finds it, whether it's a coopful of chickens (he likes eggs, too), a hutch of rabbits, a family of hatchling ducks, or a newborn lamb. The old axiom "Never set a fox to guard the henhouse" hasn't helped his reputation, and to an extent, he deserves it. But he's no rapacious killer of livestock, or anything else that isn't small and relatively weak, because he isn't all that big himself.

A North American red fox is likely to measure twenty to thirty inches from nose to rump and sport another twelve to eighteen inches of bushy tail. Males weigh in at ten to twelve pounds on the average, the typical female a pound or two less. Their central European cousins tend to grow larger, with both sexes averaging eighteen to twenty pounds.

He's a small but wonderfully capable package, possessed of all the sensory gifts devolved upon the canids—a fine nose, excellent ears, and keen eyes. Members of the genus *Vulpes* (which includes the red, swift, and kit foxes, but neither the gray nor arctic foxes) have catlike, spindle-shaped pupils; their vertically elliptical shape seems to promote extra-sharp vertical vision, which is useful to a predator.

The gray fox can climb trees almost as nimbly as a squirrel. The red fox is landbound, but it's no handicap. He's a splendid long-distance runner, covering mile upon mile in a tireless, characteristically light-footed trot; if he needs to he can reach thirty miles per hour in a sprint. He can clear a six-foot fence, swim well, and, using his tail as a counterbalance, can maneuver with the best.

◆

He's a handsome chap as well, with a long, somewhat pointy muzzle, alert-looking ears (also pointy), and a thick coat of long, soft fur. True to its name a red fox's upper body and tail generally are red, although the shade can range from yellowish to a deep red-brown. Underneath, its cheeks, throat, and belly are white, or pale, ashen gray. Legs and feet are glossy black, and the tail usually shows a few black hairs and a white tip.

Not all specimens of *Vulpes vulpes* are the classic red, not necessarily even among siblings of a given litter. Cross foxes are so called because their rusty-red coats are bisected by a black line down the middle of the back and another at right angles across the shoulders. In the rarer silver phase the coat is dark gray or black, usually frosted with white but sometimes nearly all black. Silver fox pelts are much prized in the fur trade, generally supplied by pen-reared animals commercially bred from pure strains of that particular color phase.

Geography seems to determine in large part how frequently color variations occur. Our Missouri foxes, for instance, are almost always red, with an occasional silver-phase animal mixed in. Cross foxes are extremely rare here, although they account for about a quarter of the species' total population worldwide.

The red fox's penchant for being a stay-at-home probably has much to do with it. A family's home range may be as small as 150 acres in prime habitat or as large as twenty square miles where the pickings are slim. Adjacent families do not intrude upon one another's turf, however. Even when young foxes strike off to set up housekeeping on their own they don't go any farther than they have to; usually it amounts to six or eight miles for females and perhaps twenty miles for males—all of which encourages such genetic stuff as color phases to develop regional differences.

Lifestyles by and large are not so different. Since their prey is most active at night foxes generally are, too. Having spent the day loafing in a thicket or some other protected spot a fox starts its nightly round in late afternoon, so often traveling the same routes around its home range that it creates a network of distinct trails. It seldom seems in any great rush, and if you follow a fox trail for a while you'll find places where the animal has stopped to groom its coat, leaving strands and snarls of white and rufous hair behind.

Foxes also stop now and then to roll on dead carcasses and other foul-smelling things. This, too, is a typically canine trait (the same highly bred Brittany I mentioned earlier is a real aficionado). I suspect canids do this in part at least to mask their own scent from detection by prey animals. If so, foxes need it more than some others. Like many other canids some species of fox have scent glands on the upper surface of their tails, near the base; these give off a rank, musky, slightly skunky smell that probably serves to communicate the presence of one fox to another and may even be a mark of individual identity. To those of us with less sophisticated noses, it is identifiably foxy in any case. Smell it once and you won't forget it. On still, damp days when scent hangs well, I've even found fox trails just by following my nose.

Depending upon local food supply, a fox might travel as much as five miles a night, making its circuitous way around the home range. He's a skillful hunter and can put a good variety of moves on a prospective meal. Approaching something that might see it coming—a bird or rabbit, for instance—a fox will crouch catlike, edge along low to the ground, and make the capture in a short rush. A mousing fox, on the other hand, usually stands tall, listening intently and taking high, slow steps. When the moment is right it will pounce with a high leap that seldom misses the mark, pin the mouse with its forepaws, and dispatch it with a quick bite. Foxes do a fair amount of mousing during the day and go about it so intently that you sometimes can slip up from downwind and watch for quite a while without spooking them. It's a great show.

The average fox eats a pound or so of food per day. When prey is abundant it often will kill more than it eats and cache the excess against leaner days to come. It may bury extra food or simply cover it with leaves and grass; in either case it stakes a claim by circling the site with urine and comes back from time to time for a snack or, apparently, just to check on the larder.

Red foxes generally hunt alone and aren't particularly sociable, but they do form family groups, with a male and one or two females sharing a home range throughout the year. Though foxes scarcely ever bed down in the same spot on consecutive days, each home range contains one or two dens, which may be a refurbished woodchuck or marmot burrow, an ancestral fox den, or one newly dug. In any case the vixen, the female, does the work. She'll usually choose a sunny hillside, a fencerow, some suitable spot at a woodland edge, or a cavity among rocks. Sometimes she'll locate the den under an old building or in a hollow log. It will have multiple entrances—at least two, perhaps as many as twenty-five—all of them leading to a central chamber about four feet underground. Wherever it is, the den is a key factor in the family's life, for there the young are born.

In *The Master of Game*, written about 1410 and the oldest sporting book in the English language, Edward, Duke of York, observes, "The *fixen* of the fox is a *saute* ones in the yeer." *Saute*, from the French *sauter*, means "to be in heat," and Edward was right; vixen foxes come in estrus only once a year and even then only for three or four days.

Depending upon the latitude where they live, red foxes' breeding season may come as early as December or as late as April—or as Nicholas Cox puts it in *The Gentleman's Recreation*, written in 1697, the season of the fox begins at Christmas "and lasteth till the *Annuntiation* of the *Blessed Virgin*." If there are foxes where you live you can hear the breeding season begin, because they bark and chatter more than usual—which probably is the origin of *clicketting*, a lovely term the British use to denote the mating of foxes.

Once she's gone a-clicketting (which may involve more than one male although she'll eventually form a partnership with only one) the vixen gestates her young for an average fifty-three days, at which time she may bear only a single pup or as many as ten. Sometimes two vixens will give birth in the same den, and most vixens will freely nurse any pups, their own or someone else's.

The kits, as they're sometimes called, are born blind but well-furred and weigh three or four ounces each. Their eyes open a week to ten days later. They'll remain in the den for a month or more. If the vixen leaves the den during that time, she won't go far or be gone for long. The male does the hunting, bringing food to his vixen and leaving it outside the den; if he tries to go in he'll get a nip on the nose for his trouble. She's a protective parent.

Once the pups reach an age to venture outside, it's not uncommon for the adults to move some or all of them to a second den, which may be a mile or more away. If they divide the

litter each adult takes sole charge of a few pups; if they don't the fox and vixen begin sharing the chores of finding food. She'll do most of her hunting at night and stay with the pups during the day when the male is off catching his share of the family livelihood.

Red fox pups are as playful as any young canids—or as kittens, for that matter—tumbling with one another, tossing and pouncing on discarded bones and other litter around the den. One of my uncles was a professional trapper years ago, and one spring he dug out a litter of newly weaned foxes and brought them home. I played with them for hours on end, fascinated by their endless capacity for fun. You can have a great time watching kittens play with a paper bag lying open on the floor; watching two or three fox pups do the same thing is better than television any day.

When they're about ten weeks old the young ones start hunting with their parents. At about twenty weeks, when their milk teeth have been supplanted by a more businesslike set of tools, they start hunting on their own. They leave their parents' home range in the fall, going off to set up territories of their own and, as they become sexually mature in the following spring, start the cycle all over again.

Through his long association with man the fox has become a metaphor for stealth, deviousness, cunning, and guile—sometimes more so in popular myth than in reality. A fox's tail is a useful item, but contrary to what you might hear he doesn't use it to brush away his tracks in the snow. Nor, as I once read in that continuing saga of biological half-truth, "Mark Trail," does a fox rid himself of fleas by holding a stick in his front teeth, backing slowly into a stream or pond until he's submerged to the nose and all the fleas have climbed aboard the stick, then letting it float off like Moses adrift on the Nile.

Reality is impressive enough. Unless you're using pheromone-laced vixen urine as a lure, foxes are difficult animals to trap because they're highly cautious and extremely intelligent. Certainly, they're brighter than the dogs we use in the fine old sport of fox hunting—a sport that seldom ends with a dead fox despite the weepy claptrap you might hear on the subject. If those who decry horse-and-hound fox hunting as taking unfair advantage of a defenseless animal were as smart as *Vulpes vulpes* they'd spend their time at more productive things. The fact is, a red fox can lead a pack of dogs on a merry chase, backtracking to confuse the scent, swimming to interrupt it, running in intersecting circles, generally raining chaos on their heads, and behaving all the while as if he enjoyed the game. If he goes to ground he's out of reach, and any fox dumb enough to turn and fight a pack of dogs doesn't belong in the vulpine gene pool anyway.

Neither swift nor kit foxes are as wary or as elusive as grays or reds, and their numbers have been reduced in some areas by predator-control campaigns of poisoning—which is to my mind a reprehensible and cowardly thing. Habitat destruction is a factor in some areas, as it is for most wildlife, but foxes are adaptable enough to maintain their niche almost as fast as we can change the rules. Indeed, the last thing the last human on earth sees may well be the face of a red fox. Smiling.

◆

# II

# Charm and Character

# 6

# Gentleman of the Fields

Ask me to name the first bird call I ever heard, and I couldn't begin to tell you what it was. But I certainly can tell you the first one I could identify, because it came from a little gentleman whose favorite song is his own name.

I was a seasoned woodsman of eight or nine before I could be certain of a cardinal's song or a chickadee's or even the blue jay's rowdy clamor, but I cannot remember a time when I had to wonder who was the source of a sweet, clear April voice spilling *ah-bob-white...bob-white...ah-bob-white* over and over into the still, cool air of early morning.

Apart from starlings and pigeons, my father dearly loved anything with feathers and wings, and loved the sounds they made. He passed the same affection on to me, both for birds in general and for the one that always was his special favorite. If I care more for woodcock than he did, it's only because I've spent more time with them and because their strangeness and mysteries tease my imagination as no other does. But for all that, no bird rings in me a deeper note of homely charm than quail.

Specifically the bobwhite quail, *Colinus virginianus*, peerless dweller of fencerows and farmland edge, six or seven ounces of sheer grit and quiet determination. Like any gentleman, he dresses well—in gray and copper, reddish-bronze and black, with a formal touch of white at his throat—and he sports a pedigree that reaches back a million years or more.

His relatives, near and distant, are legion, for quail belong to the huge order of gallinaceous birds whose members include turkeys, pheasants, grouse, partridge, ptarmigans, and others—an order that comprises a host of species worldwide. Both Old and New World quail share the subfamily Phasianinae with pheasants and partridge. Only the relatively small tribe of American quail, about thirty-six species, are native to the Western Hemisphere; of these only a half-dozen belong to the northern continent, and only one is truly widespread.

The bobwhite originally ranged virtually throughout the eastern United States, from Wyoming to southern Ontario and Massachusetts, south to the gulf, through Texas and all along the eastern side of Mexico to Guatemala. It has since been introduced in areas of western Mexico and the Pacific Northwest.

All other North American quail belong solely to the West. The blue or scaled quail is the only one whose range overlaps the bobwhite's original range. (This bird is named for the lovely gray-blue color of its feathers and for the scalelike pattern on its breast; it's also called cottontop because the feathers of its crest are tipped with white.) California and mountain quail are largely restricted to the Pacific Coast. Gambel's quail, whose plumage is very similar to the California quail's, inhabits the Southwest and parts of western Mexico. So does the distinctive Montezuma quail, whose black and white harlequin-patched head markings, white-speckled flanks, and stubby, almost nonexistent tail make it impossible to mistake for any other bird.

◆

Where ranges adjoin or overlap, which species you're likely to see depends in some measure upon the altitude where you look. Mountain quail tend to seek the highest elevations, although you won't find them above timberline. California quail prefer somewhat lower altitudes, which is why they're also known as valley quail. Gambel's quail like lowland deserts.

Wherever they live and whatever the species, quail are quail—ground-dwelling and gregarious. A typical covey of midwestern bobwhites contains about fifteen birds, but late in the year smaller coveys sometimes combine. I've seen sixty- and seventy-bird coveys in the Mexican state of Tamaulipas, where you'll find the greatest population of bobwhites in all the world. Blue quail often gather in groups as large as a hundred, which I suppose is more properly called a flock than a covey. And I've been told by people whose word I trust that certain valleys in the Baja Peninsula are home to California quail in congregations of a thousand birds. I wouldn't know what to call a group that large—except perhaps The Quail Hunter's Vision of Heaven. I intend to go see it for myself someday.

Birds that spend most of their time on the ground need good means of self-preservation. Some have evolved a hair-trigger approach and flush at the merest hint of danger, but the majority prefer either to run or to hide. Turkeys and pheasants are great runners, for instance, while woodcock and ruffed grouse are more likely to sit tight and simply blend into the background. Quail do both.

Bobwhites and Montezuma quail are experts in the game of freeze-and-fade. Of all the birds I know, only the woodcock is inclined to sit tighter than a bobwhite, which is why quail hunting in its classic form is the milieu of pointing dogs and small-bore guns. But Gentleman Bob has an answer for that, too; a quail holding motionless often compresses its feathers so closely that its eyes seem to bug out. I take this to be an instinctive means of reducing body scent.

◆

In areas where hunting pressure is relatively high—or if a covey is surprised in thin, open cover—bobwhites are more prone to run than hide. Even so, they can scarcely hold a candle to the western species. Blue, mountain, California, and Gambel's quail are all hot-footed little chaps whose legs turn to blurs in response to a threat. Their plump bodies and erect posture make them look like mechanical toys in hyperdrive, but it's a splendid defense, for their speed is remarkable and they invariably head into the densest cover they can find. A man hasn't a chance of keeping up even where the habitat is merely thick; when it comprises the sort of vegetation common in the Southwest and Mexico, trying to flush a covey is damn near an exercise in suicide.

There, in the mesquite wilderness known locally as *brasadera*, or black brush, you're confronted with an impenetrable mélange of prickly pear, Spanish dagger, devil's head, *coma, brasil, clepino, retama, granjeno, tasajillo,* and about twelve zillion more, all bristling with spines, spikes, needles, hooks, thorns, and every other kind of botanical weaponry. Having spent a fair amount of time hunting cottontops and bobwhites in this sort of stuff, I can tell you that all the advantages belong to the birds. But then no bird hunter worth the doghair in his drink would have it any other way.

If legwork or camouflage are a quail's first line of defense, flight comes a close second. Every species is equipped with the short, broad wings of an aerial sprinter, and I don't know of any that aren't capable of hitting top speed in three or four strokes. That their flat-out rate may be only about thirty-five miles per hour places them well down on the list of avian speed-demons, but it's fast enough, and when you add in the chaos of ten or twenty or a hundred little bodies all hurtling out at once, it's clear that quail have refined the notion of safety in numbers to a fine art.

This is especially true of bobwhites in their winter coveys. These are superbly organized units under no particular leadership. There are no alpha quail of the pack, no matriarchs of the herd, no individual from whom others take their cues. A covey of bobwhites is simply a congregation of individuals who do everything together. They forage, loaf, and roost as a unit, and at times they behave with the precision of a single organism.

One of the most touching demonstrations of this is the way a covey sleeps. A mated pair or family group—which, as I'll get to presently, is not a true covey—will occasionally roost in trees or low bushes (California quail do this, too), but a covey roosts on the ground, more or less in the open. In late afternoon they move out of whatever cover they're using as a loafing ground, feed a bit and gravitate toward a stand of grass or grain stubble or a weedpatch. There, in the last fading light of sundown they form up in a circle, sides touching, heads pointing outward all around. On a cold night they'll snuggle especially close to share their warmth, but they always sleep sharing touch, and if they're startled or attacked, the sudden movement of one bird communicates instantly among all the rest and the whole group explodes into flight.

You can recognize these roosting sites as circular patches of grayish-white droppings. The diameter, ranging from teacup to dinner-plate size, will give you a rough idea of how many birds make up the covey. (Blundering into a roosting covey just at dark will also give you a rough idea of how heart failure must feel. A covey flush at close quarters always startles me, even when I'm expecting it; taken unawares, it's like sticking your finger into a lamp socket when you didn't know the juice was on—except mere electrocution isn't accompanied by the sudden roar of wings or the pandemonium of pudgy little birds flailing around your head like a swarm of bees.)

Finding one roost, the chances are you'll find others nearby, but this doesn't necessarily signal the presence of other coveys. Left undisturbed, bobwhites aren't inclined to travel far. A covey may roost in one small area for several nights, perhaps for as long as a couple of weeks. And they're likely to follow a uniform schedule, often showing up at the same places at about the same times each day.

◆

The degree of coordination the covey shows is all the more remarkable for the fact that it's not a long-term association. When winter loosens its grip—as early as February or as late as April, depending on the latitude—the coveys break up, and not just temporarily. If even two or three birds share the same group the next time coveys form, it will be purely by accident.

Coveys dissolve for the same reason they formed in the first place—survival. In spring, however, the imperative is directed not toward the current generation but rather toward the next.

Just before the winter group breaks up, the cocks grow peckish with one another, ruffling and bluffing and swaggering as self-importantly as if they swung many times their actual weight. Finally they begin the lovely ritual of advertising. From a low branch or a fencepost or a mound in some open place, each proclaims his personal turf and his intentions. Ruffling up to half again his real size, he suddenly stretches tall, opens his stout little beak, and with a singular lack of modesty shouts his name at the top of his voice—*ah-bob-white...ah-bob-white...ah-bob-white*. Sometimes he seems to drop the first syllable, and occasionally he might stick in a single *white* as if running out of time or breath.

It's all to show the ladies what a handsome fellow he is. For the most part, pair bonds are formed by the time the covey breaks up. Some cocks, apparently unable to find mates, continue calling well into the nesting season, and even a few well-mated chaps go on as well, as if they're just too wound up to quit. My old friend Jack Stanford, for many years the country's preeminent quail biologist, calls them "psychological bachelors." Having known a few people with similar tendencies, the description makes sense to me.

The odd Lothario aside, however, bobwhite cocks generally are first-rate family men. In fact, they do most of the nest-building work. A stand of grass is their favorite sort of nesting site, especially if it's one of the clump-growing species like orchard grass or redtop or bluestem. Finding a bare spot, the cock scratches out a shallow, bowl-like depression. The hen tinkers with it a bit, forming some overhanging grass stems into a canopy, but she leaves the main jobs of shaping and lining the nest to him.

Once it's finished to her satisfaction, the pair play out their mating ritual, which involves a good deal of strutting and stretching, tail-fanning, wing-shaking, foot-stamping and other posturings on the cock's part and, when she's convinced, some wing-fluttering and tail-twitching from the hen in response. They'll go through the same routine at least once a day for better than two weeks, and most days the hen adds another egg to the nest until she has a full complement of a dozen to fifteen.

When the last egg is laid she rolls and turns them all until they're neatly arranged and only then settles in to begin the twenty-three-day session of incubating. She'll take short breaks now and then to feed, but otherwise she sticks faithfully to the job, turning her eggs several times every day to make sure they stay uniformly warm. If something happens to her, if she's snatched by a hawk or a fox or a weasel or a foraging cat, the cock will take over and finish brooding the clutch. But as long as she's okay he spends his time loafing nearby and calling frequently to let all the other cocks know that the territory is occupied. Should he fall victim to predator or mischance, the hen may later take up with one of the neighborhood's unmated cocks. The nest itself may be wiped out by heavy rain, fire, mowing machine, blacksnake, raccoon, or some other agent, and if it is the pair will start over again in another location.

At precisely three weeks' time, each chick pips a tiny hole in its eggshell, but they'll need two more days to complete the task of chipping most of the way around the large end so they can push their way out. Once they reach this point the whole brood will hatch within about three hours.

A just-hatched bobwhite is a tiny, damp, bedraggled-looking little customer scarcely able to lift its head. Its eyes are tightly shut and its natal down looks more like fine hair than feathers. But after a couple of hours under the hen's warmth, bobwhites turn to wide-eyed, fluffy little balls of gray- and buff-colored fuzz atop spindly legs and ungainly-looking feet. As soon as the hen steps off the nest, they venture out as well—not far at first, and they're soon back, huddled beneath their mother. Before the day is out, however, the whole family has abandoned the nest, the chicks scrambling about, exploring and snatching at insects, both parents riding herd. From then until they're too large to cover, they'll spend each night in the shelter of their parents' bodies and wings.

They seem such delicate, vulnerable little sprouts that it's hard to imagine them surviving, and indeed, they're beset by all sorts of hazards from hailstorms and chilling rain to predatory animals. But they are astonishingly capable. For several years when I was a wee lad, quail nested in a certain corner of my grandparents' garden. Granddad always knew when the eggs were about to hatch and made sure I was there to see it. Then he'd announce that beginning on the third day hence he would pay a bounty of one buffalo nickel for every chick I could catch by hand and present unharmed. He had, he always said, a great tooth for quail on toast.

I thought it a handsome offer, though he knew full well that neither his supposed appetite nor my piggy bank were due for much satisfaction. It's a pity there were no video cameras in those days, because a tousle-headed kid scrambling madly under the grape arbor must have been a funny sight.

I might as well have been trying to snag gnats or can a will-o'-the-wisp. One moment they were there, the next they were gone. An old folktale has it that quail hide by lying on their backs and holding leaves in their beaks for concealment. I can tell you it isn't true, but I can also tell you they don't need any such chicanery, not even days-old chicks. All they have to do is squat down and they're as well camouflaged as any bird needs to be. And when they decide to split, they can run like blazes. A dog or cat can catch one, sometimes, but an eight-year-old boy doesn't stand a chance.

Like all gallinaceous birds, adult quail are seed-eaters, but insects are vital to the young. Fueled by the rich animal protein, they're the size of sparrows at the age of two weeks and can fly short distances. At four weeks their bodies are as large as a robin's. By then they still wear a few wisps of natal down, but they're almost fully dressed in nondescript coats of juvenal plumage. At the end of ten weeks they're the size of their parents, and a couple of weeks later you can begin to tell the cocks from the hens. They're fully grown fifteen weeks after hatching.

In the central latitudes, hatching is at its peak from the middle of June through the middle of August, but a few mated pairs, dogged by the loss of several nests and clutches, may still be laying eggs. Chicks hatched very late generally don't survive, since the weather turns cold before they're old enough to cope with it, but even so, the adults' persistence is a good thing, because bobwhites need the strength of numbers for winter survival.

None of the American quail are migratory, but some Old World species are—notably *Coturnix coturnix*, a Eurasian bird historically known to travel in vast numbers across the Mediterranean and the Red Sea. This undoubtedly is the bird referred to in the Book of Numbers: "And there went forth a wind from the Lord, and brought quails from the sea...And the people...gathered the quails: he that gathered least gathered ten homers...." As the homer was roughly equivalent to ten and a half bushels, scriptural scholars estimate that this gathering, performed by 3 million Israelites over the course of two full days and a night, accounted for more than 20 billion birds.

This may be metaphoric rather than literal reporting, but there's no question that C. *coturnix* once was mightily abundant. Pliny describes them arriving in Italy so exhausted from flight that they often settled aboard offshore boats for a rest—and in such numbers that said boats promptly sank under the weight.

As I said, American quail do not migrate. Instead, they form coveys and begin to practice the close-order living that helps ensure a high rate of survival through all but the harshest of winter conditions. To help ensure the most capable coveys, nature has evolved in quail a brief but vitally important period of apparently chaotic behavior.

It begins about the time the first frosts start to thin out the once-lush vegetation. Through late summer and early fall, the birds have lived as family groups, staying mostly within their own territories. Now, however, they grow restless. The families break up and scatter in all directions, sometimes traveling considerable distance. This is the equivalent of mad-moon behavior among grouse. The old-timers called it the fall running; quail biologists nowadays prefer to call it the fall shuffle, and that's exactly what it is.

For solitary-living ruffed grouse, the autumn diaspora is an attempt by newly grown birds to find territories of their own. For quail it is the means by which winter coveys form. The shuffle once was thought to be nature's way of guarding against inbreeding, and you can still find some old-time quail hunters who cling to the belief. The fact is, though, researchers have found no predictable or consistent ill effects from inbreeding, even among carefully segregated populations. Sound stock produces sound offspring, and if some genetic glitch happens to show up, those birds simply become part of the natural mortality cycle.

What the shuffle does accomplish is far more important than short-circuiting incest. When the running is over, bobwhites of varying ages are thoroughly mixed, and any winter covey not composed entirely of very young or relatively old birds stands the best chance of survival. Moreover, the shuffle itself helps cull out the weaker individuals. During this period quail show up in all sorts of odd places—city parks, suburban lawns, golf courses, barnlots—places where they're vulnerable to predation and mishap, which tends to leave only the fittest and most fortunate to make up the coveys.

Unfortunately, though, not even the fittest covey can live where no habitat exists, and over the past forty years quail habitat has disappeared at a discouraging pace. Bobwhites don't require all that much, but what they need they need absolutely—

good nesting grounds and hard, brushy cover with a diversity of plant life nearby. Since the 1950s the trends of American agriculture have gone precisely the opposite direction. Hedgerows, fence-corners, and ditches have been bulldozed by the thousand; grainfields are planted to every available inch, drenched in herbicides, and the leavings plowed under every fall.

Worst of all, millions of acres that once grew native grasses and annual weeds now are biological deserts of fescue—a hardy, aggressive, sod-forming, thrice-damned imported grass once hailed as the cattleman's salvation. In fact, the lousy stuff is about as nutritious for cattle as broomstraw and offers even less for wildlife. A cottontail can starve on a fescue diet. Quail can't successfully nest in it because it grows so densely at ground level that chicks can't move, and if they can't move, they can't feed. To anyone who loves wild things, *fescue* should be the filthiest f-word there is.

Yet in spite of it all bobwhite survives, even to abundance in many places. And while he does, the portion of my heart that belongs to him remains intact. I couldn't begin to total the miles I've traveled to spend time in the company of quail, nor the hours that time would amount to, but every minute of it has nurtured something in my soul. It's gone on for nearly fifty years, and yet it never fails to somehow lead me home—whether listening to their springtime songs, watching the families forage and dust, tramping sere brown fields to the sound of dog bells, or simply marveling at how wonderfully resilient they are.

Bobwhites make a particular call when a covey is scattered and they're seeking to locate one another and get together once again. Learn to imitate it, sit still, and you can summon the birds right to you. It's a sweet, soft, two-syllable whistle played on a rising scale. It sounds like a question, and it's filled with answers. To me, the covey call transcends time and space, spans the gulf that separates the living and the dead, spins forward to tell of cycles yet to come. I would not be at all unhappy if it were the last sound I ever heard.

◆

# 7

# Little Ghost of October

"And where is *yoouur* favorite place in the world to hunt, Mr. McIntire?" the lady asked, lending the tones of a duchess to our cocktail-party chat. She and her husband were just back from Tibet—which for some reason she chose to pronounce with the accent on the first syllable. They were a pleasant, elderly West Coast couple who amused themselves by making travel films organized around his enthusiasm for big-game hunting.

By way of reply I described a certain forty-acre alder swamp in northern Minnesota.

"And what do you *huunnt* there?"

"Woodcock."

She paused a moment looking politely puzzled and then slowly said, "Woodpeckers...?"

"No. Woodcock. Sometimes it's called 'timberdoodle.' It's a six-ounce bird with a three-inch bill, eyes on top of its head, ears next to its beak, and a brain that's upside down. It eats earthworms."

She looked as if I had spoken in Urdu or suggested we go for a skinny-dip in the hotel pool, then turned to her husband, who was talking with someone else.

"Philip. Philip. Do we know anything about...wood...cock?" she asked, sounding like Margaret Thatcher being forced to say "toilet paper."

"We don't have them in California, dear," he said gently.

"Oh. Well. Of course," she said, turning back to me as if that cleared up everything. I considered telling her there were no woodcock in *TIBet* either, but my mother always told me not to tease my elders. I could've got away with it scot-free though, because all evening long she clung to a firm conviction that I was someone named Miles McIntire.

Though the good lady's reaction certainly had its quirks, it really wasn't unusual. I've seen the same puzzlement on faces belonging to people who have lived a lifetime around prime woodcock habitat. Describe woodcock to those who've never seen one and they think you're putting them on. Show them one and they still think so. Fact is, a woodcock looks like something designed by a committee and assembled in the dark by a mad mechanic using leftover and apparently unrelated parts.

But a woodcock is the most utterly charming, thoroughly fascinating bird on earth, or it is to me anyway. It was love at first sight, and in these forty-odd years since my father brought home one he'd shot while hunting quail the feeling hasn't changed, except to grow stronger.

Arguably any animal so weird-looking almost *has* to be lovable, but there's more to it. If anything about woodcock is commonplace, mundane, or uninteresting, I haven't found it.

Their ancestry, for instance. By pedigree a woodcock is a shorebird, related to sandpipers, curlews, godwits, snipe, turnstones, phalaropes, yellowlegs, dowitchers, and all the roughly ninety species of ground-dwelling wading birds that make up the family Scolopacidae. The woodcocks themselves belong to the genus *Scolopax* and not surprisingly account for only a small part of the family. *Scolopax rusticola*, the European woodcock, is the largest of them and also the most widespread. It lives virtually throughout Europe and northern Asia and even in some high-altitude regions of Indonesia and New Guinea. Two other species live in Celebes and the Moluccas. The American woodcock inhabits all of North America east of the Great Plains, from southeastern Manitoba through the Maritimes and south to the Gulf Coast. Ornithologists now know it as *Scolopax minor;* it used to be *Philohela minor*, "little bog-lover."

Our bird is considerably smaller than its Eurasian cousin, tends more toward russet-brown than rusty-red and also lacks *rusticola*'s distinctive barring on the throat and breast. Structurally though, a woodcock is a woodcock—and therefore is unlike any other bird alive.

Long, pointy wings, short tails, slender bills, and a fairly streamlined shape are typical of nearly all the sandpiper kin. Woodcock have kept the family wings and tail and bill shape, but over the ages since they turned renegade and abandoned a wading, shoreline life in favor of the uplands, they have evolved short legs, short necks, and charmingly pudgy bodies. The most distinctive changes, however, have occurred from the shoulders up.

In nature, form and function endlessly shape one another. Shorebirds obtain their food in or near water, tweezing up tidbits from the mud or sand or shingle. The longer their legs and bills, the deeper the water they can exploit as a feeding niche. Woodcock, on the other hand, make their living by probing soft, moist soils for earthworms. Long legs would be a handicap, although they don't mind wading into an inch or so of water, but the bill needs length and also some special capabilities for working at full reach underground.

The top mandible of a woodcock's bill is flexible at the tip—prehensile actually, so the bird can open the end of its beak and grasp a worm even with its bill plunged nostril-deep into the earth. The upper mandible also is lined with tiny, backward-pointing, toothlike structures, and its tongue, roughly the same length as the bill, is raspy—all to provide purchase on a worm's slippery, muscular body.

Earthworms make up about 90 percent of their diet; grubs, insect larvae, snails, and the odd moth account for most of the rest. Consequently a woodcock's digestive system has little need for a grinding chamber so the birds have no gizzards, only an esophagus, a stomach, and a long, spirally coiled intestine. In Europe woodcock traditionally are cooked with the innards left in, and the intestine, called the "trail," is thought to be a great delicacy. I think so, too.

Because woodcock feed mainly at dusk and at night, they have big, light-gathering eyes. Relative to body size, woodcock eyes are enormous, as large proportionately as an owl's—deep liquid black pupils surrounded by dark-brown irises that show only in the brightest light. A man can get lost in those eyes.

Very little, however, is lost to them. A fair number of predators share the woodcock's penchant for a crepuscular and nocturnal lifestyle, and a bird that dines with its nose in the mud has good use for eyes in the back of its head—or on top, which is better yet. In the process of refining its adaptations the woodcock's eyes have migrated to the top of its skull, literally behind the ears, the openings of which are low down, near the base of the bill. As a result woodcock possess a full 360-degree field of view with zones of binocular vision both before and behind. The backward zone seems to be slightly wider than the one in front, which probably accounts for the bird's demure-looking, bill-down posture both on the ground and in flight.

As the eyes moved up, the brain shifted position—shifted so radically in fact that it turned completely over. A woodcock's brain stem now is on top, with the larger lobes underneath. The inversion served two purposes. For one it allowed more room in the cranial dome, thereby accommodating higher placement of the eyes without requiring the skull to be any larger. More important, it also allowed the cerebellum to grow proportionately larger than that of most other birds, and because the cerebellum controls motor skills and coordination, this in turn enhances the woodcock's maneuverability on the wing.

Which is no small gift considering where they live. You'll find the occasional straggler out in the open, but woodcock mainly prefer habitat that offers protection overhead and relatively easy walking underneath—which also is habitat that demands some sophisticated flying skills from a ground-living bird. Going from branch to branch is one thing, and so is blazing off at low altitude

in high-speed, curving arcs as grouse do; rising in full flight through three dimensions of tangled overgrowth is quite another thing altogether. Woodcock don't come even close to matching the astonishing aerobatics that hummingbirds perform, and they certainly aren't as powerful as grouse. But they can weave a quick, nimble path through branches and treetops that scarcely any other bird could follow.

◆

Form and function. Woodcock are superbly maneuverable; woodcock habitat requires superb maneuverability. This is true of both the upland loafing cover where they spend the daylight hours and of the moist feeding coverts they head for in the evening—making what is traditionally called the dusking flight. In either case, habitat is one reason why the woodcock is an unfamiliar bird to so many people. Woodcock don't come to us. We have to go to them, and where they hang out are places most wouldn't choose for a pleasant stroll. It might be an aspen thicket, young alders, cedars, hardwoods, sycamore jungles, or willow brakes. It might be brushy, middle-aged timber or regrowth no more than head-high to a man, but so long as it forms a canopy and is not choked with grass it's likely to look good to a woodcock.

Assuming, that is, it looked good last year and looked good to the preceding generations. One of the still-unrevealed mysteries of woodcock everywhere is their uncanny persistence in choosing the same small patches of habitat year after year. To a hunter a woodcock covert is a secret to cherish, because it will continue to be a woodcock covert for years to come. No one knows nor probably ever will know why they like this tiny corner of the woods and not that one, why this stream bank and not the opposite side. Only the birds know why. All we know is that at certain times of the year they visit certain places, linger a while, and then they're gone.

Or at least that's how it is in the upper reaches of woodcock range, north of a line curving from the northwest corner of Louisiana across to the Carolinas and up the Atlantic Coast to New Jersey. A small population remains in the southeastern United States year-round; the rest migrate up and back, retracing some genetic map that seems to return each bird to where it was hatched, however near or far the place may be.

Spring means north again, and even if spring comes late the woodcock might not. I've seen them in my overgrown pastures here in the middle of Missouri as early as the third week of February, noodling around among patches of grainy snow, and have watched many times at sundown while the little sky-dancers perform.

Having seen the mating rituals of everything from yuppies to scorpions, I can only say that none seems more ethereal or more beautiful than the woodcock's elegant gavotte. It is a solo act for the males, meant to impress females that are larger by a third, but if the little chaps are intimidated courting lady-loves of Amazon proportion, they don't show it. When the light level is just right, usually somewhere between 5 and .02 foot-candles, they take center stage on the singing ground. This itself is an ancestral venue where they or their fathers or grandfathers have wooed and wed before. Why this particular place and not another? You'll have to ask the woodcock.

As a preliminary they ruffle their feathers and turn in tight circles with their tails fanned like miniature turkey-cocks, tail-feathers showing brilliant white tips on the underside. And then they "peent" and sing.

Under most circumstances and during most times of year woodcock are not very vocal. Hens use a small repertoire of quiet sounds to communicate with their chicks, but otherwise the most common sound you'll hear from a woodcock is the soft, eerie twitter their wingfeathers make in flight—except on the springtime courting grounds. Then it's as if Harpo Marx suddenly becomes a balladeer.

The warmup is customarily called "peenting," although *peent* is hardly an accurate rendition. It sounds more like *bzzzzt!*, a quick, insectile note of about a quarter-second duration. On a sonogram it covers nearly four octaves above middle C. In homier terms it sounds like a cicada trying to kick-start its sawlike buzz; so next time you hear a cicada in early spring, well before cicadas begin to sing, what you're really hearing is a male woodcock peenting. Follow the sound through the gathering dusk—it carries a quarter mile or more—and if you approach quietly enough you can watch and listen to one of the finest shows of spring.

◆

I suspect peenting serves as a come-on to attract the attention of any females nearby and also to warn off other males. How long it lasts seems to depend entirely upon the singer's whim. I've seen them begin their courtship flight after peenting only a half-dozen times. But I once spent a damnably uncomfortable twenty minutes hunkered motionless under a cedar tree while one long-winded little guy stood six feet away, peented 138 times (I kept count), and never did perform. He just walked away, leaving me to unhinge my knee joints as best I could.

More often, though, he peents a few times and then takes off in a rising spiral, wings twittering, flying round and round the singing ground, climbing steadily into the darkening sky. He will reach two hundred feet or more, until he is the merest speck or lost to sight altogether. At the apex he pauses for an instant and then drops, fluttering, sideslipping like a leaf, spiraling and spinning, all the while singing an incredibly lovely, trilling, liquid melody of kisslike notes. It fills the sky as if coming from everywhere at once. A few feet above the earth, just as he seems sure to crash, he brakes his descent, settles back where he started, ruffles, struts, peents, and plays the whole scene over again.

The display may continue for half an hour or more, considerably more if the sky is clear and the moon full. I don't know what your ideas of heaven are, but one of mine looks a lot like a woodcock singing ground on a moony spring night.

I have never been able to spot a female on a singing ground, but if I'm close enough to accommodate my failing ears I know she's there; I can hear the soft little gurgling note the male uses to precede his peents when he knows a female is present. I believe she utters it as well on occasion. William Sheldon, the dean of woodcock researchers, describes the sound as *tuko*; to my ear there's an *l* in it somewhere.

The whole display presumably impresses her at least as much as it impresses me, since the singing ground is also the mating ground. And the nesting ground won't be far away, anywhere from a few feet to a few hundred yards.

◆

While courtship is strictly a male gig, nesting and brood-rearing belong solely to the ladies. Once the tryst is complete he's outta there, back to his solitary lifestyle interrupted only by equally brief liaisons with other such females as he's able to tempt. He's such a lusty wee fellow in fact that researchers and prurient-minded graduate students frequently use taxidermy-specimen female woodcock as a means of observing mating behavior. Sheldon reports finding one male so especially randy that he and his colleagues were moved to name him Whorehouse Harry.

For her part the female seems not to mind in the least. She spends a few days finding a spot that suits her—typically underneath a bush at the edge of an opening in the woods—and working a little depression into the forest-floor duff. There she will lay four eggs over the course of about five days and will brood them for three weeks, feeding at night and spending her days on the nest.

As their primary means of protection woodcock rely on the natural camouflage of their plumage—cryptic coloration, zoologists call it—but none do so with more absolute confidence than brooding hens. If you can find her, which is virtually impossible except by sheer accident or with the assistance of a pointing dog, you can sometimes pet her without prompting a flush. I wish I could tell you how it feels to reach slowly down and stroke a mother woodcock's head while she sits there still as a stone, her lovely eyes unblinking. I've done it; I just can't describe the feeling—except I imagine it's like laying your finger on the hand of God.

The payoff for her anonymity and devotion to the eggs comes as an extremely high rate of nesting success, higher than that of any other ground-dwelling game bird—or of most birds in general, for that matter. The chicks hatch at twenty or twenty-one days, splitting the eggshells lengthwise, which is one more way in which woodcock are unusual. They leave the nest as soon as their down dries and begin foraging for insects.

◆

A woodcock chick comes out of the egg sporting a bill fourteen millimeters long and a pair of feet nearly as large as its mother's—so big in fact that researchers use the same size leg band for adults and young alike. They're gawky little customers and great fun to watch if you can find them; their buffy, dark-streaked natal down is splendid camouflage, and having inherited their parents' knack for squatting motionless, they can literally hide in plain sight among the leaves. Should you happen onto a woodcock brood, be *very* careful where you step and don't try to move in for a closer look. If you stay quiet and keep some distance you might get to watch the hen herd her little charges out of harm's way.

Among all the great and long-standing mysteries surrounding woodcock, none is more intriguing than the matter of hens carrying their young in flight. For at least two hundred years some naturalists have insisted that females carry their chicks—held in their bills, between their bills and breasts, clutched in their feet, grasped between their legs, even on their backs. Others have insisted just as strongly that they do not. Traditionally the scientific community has largely discounted the notion. Females flushed from a nest do sometimes fly with their tailfeathers tucked down in an odd way, and skeptical ornithologists cite this as the source of the chick-carrying myth.

Agreement is by no means universal, but current evidence suggests that it may not be a myth after all. A hen woodcock certainly *could* carry one of her hatchlings in flight—not in her bill or feet, because woodcock bills and feet aren't made for grasping, and not on her back either, for much the same reason. But she could hold one between her legs. Whether she does is another question. So far as I know no hen has ever been photographed carrying a chick, and I've never seen it happen. But some of the scientific literature says that chick-carrying has been "recorded" and "authenticated" among both the American and European species. So, maybe. I'm not willing to wholly discount any strange thing a woodcock might do.

◆

Regardless of whether she carries them, she certainly takes good care of them. The survival rate for hatchlings is about as high as the hatch rate, owing in part to their cryptic markings, in part to brood size, which is small enough that the hen can look after them all, and in part by their rapid growth.

By four weeks they're nearly full grown and able to fly, and they strike off on their own shortly after. They begin probing for worms within just a couple of days after hatching, each day using a bill two millimeters longer than it was the day before. In fact you can accurately age a woodcock chick during the first two weeks of its life simply by measuring the bill, subtracting the fourteen millimeters it started with, and dividing the remainder by two. The number you get is the bird's age in days.

Even after they reach full size you can still distinguish an immature or juvenile woodcock from an adult by the markings on its secondary wingfeathers. Young birds show well-defined dark bands near the tips of these; older ones do not. Once they're dressed in adult plumage, however, age is difficult to determine.

So is sex, for that matter, at least in some cases. Females are noticeably larger than males as a rule, but there's naturally some overlap of small hens and large males. Depending upon where you are, the time of year when you see migrant woodcock has some bearing on which sex you're most likely to find. Females predominate in the early flights. Males wait a bit longer to make their autumn trek.

Migration, too, has its mysteries and wonders. From an engineering standpoint woodcock are unlikely long-distance travelers. They are not power cruisers in a class with geese and ducks, nor slender, gliding aircraft like terns. They can cover as much as a hundred miles in a night when conditions are right, but even so, woodcock migration is a series of short hops along chains of feeding and resting coverts.

For the most part these ancestral routes are north-south river valleys, which the birds travel at altitudes low enough to make navigation a simple matter. Evidence suggests that some have learned to follow certain north-south highways as well; under moon- and starlight the concrete ribbons look something like water, and their margins frequently offer tiny pockets of habitat the birds can use.

Wind is a key factor. When the lowering angle of the autumn sun tells woodcock it's time to quit the north country, prevailing winds shift into the northern quadrant, and nothing wafts a southbound traveler along its way like a tail wind. If the breeze should temporarily shift toward the east or west, they drift along with it, using habitat along tributaries of the river systems they follow. A south wind in October is the woodcock hunter's friend, for they won't fly against it and instead stack up in the coverts, sometimes in astonishing numbers.

Woodcock migrate alone, just as they do nearly everything else in their lives. But because they are location-specific they tend to concentrate in certain places. The northern Minnesota alder swamp that started the whole thing with Her Ladyship years ago is one of them. After a few days of south wind I have moved as many as seventy-five birds there in less than two hours' flogging through the brush, and I have gone back the following morning to find the place empty, simply because the wind shifted during the night. Such numbers all at once are rare though, and farther south, where prime habitat comes in smaller pieces, the birds customarily trickle through the coverts a few at a time.

Woodcock use two principal flyways in North America—one along the East Coast and the other through the Midwest. Both culminate in the Deep South, from Louisiana to Georgia. And in both flyways there are clear signs that all is not rosy.

According to U.S. Fish & Wildlife Service estimates, woodcock populations have declined in the past twenty years—by more than 35 percent in the East and nearly 20 percent in the Midwest. *Estimate* is the crucial word, however. Studying woodcock is like studying foxfire. Even after two generations of serious scientific attention we still don't have a reliable ballpark notion of how many woodcock exist in North America nor what effect environmental degradation is having on the population at large. Under the circumstances, statistics can only suggest that something is wrong, not how wrong it might really be.

Habitat loss certainly is a problem, particularly along the traditional migratory routes. Ironically, a preservationist attitude toward woodland ultimately is inimical to woodcock because overly mature forest offers the birds little or nothing by way of food and cover. Like most woodland wildlife, woodcock thrive where forest is in the process of renewing itself, either as the result of cutting or fire, which is nature's oldest management tool.

Water pollution, loss of wetlands, acid rain, pesticide residues accumulated in earthworms—all these and more—surely affect woodcock populations. But we don't know exactly how nor to what extent. There's no evidence that hunting pressure has any adverse impact as seasons and bag limits are currently regulated, but there also is no evidence that it isn't harming the population. Too many questions, not enough answers.

In a long-overdue effort to correct these deficiencies, the U.S. Fish & Wildlife Service in 1990 completed a new American Woodcock Management Plan and entered into an agreement with the U.S. Forest Service and The Ruffed Grouse Society under which all three will cooperate in a conservation program to restore, improve, and manage woodcock habitat on public land and encourage private landowners to do the same. The plan also calls for greater effort in gathering reliable data on woodcock populations and how human activity affects them.

It's a start. How well it works once put into practice remains to be seen, but any intelligent, active effort on behalf of woodcock deserves support.

I will leave the matter of scientific data to the scientists, but in the meantime I have a few nonscientific mysteries of my own to ponder. Why, for instance, some bird dogs refuse to pick up dead woodcock. A few even refuse to point them. Do they smell bad? They don't to me, but who am I to argue nasal acuity with a dog? Fortunately my Brittany girl, October, thinks hunting woodcock is more fun than break dancing in cowflops, but I'd still like to know why some dogs don't.

I'd also like to know how a young fall bird, traveling alone on a trip it's never made before, knows where to find precisely the same tiny coverts its ancestors have used for years. And how will it find its way next spring to the singing ground where its parents met?

How do they locate earthworms underground? The tips of their bills are supplied with nerve endings sensitive to vibration, but can woodcock hear worms as well? Did their ears move forward for some reason in addition to making room for their eyes?

And just how does the world appear to an animal whose brain is upside down, anyway? A couple of my friends who consider me borderline loony when it comes to woodcock have suggested that my own brain is starting to tip, so I may eventually learn the answer to this one.

Or maybe I won't, and it won't matter a bit, not as long as I can see the little sky-dancers in the spring, marvel at the occasional brood of chicks I find on the farm, relish the moment between dog bells falling silent and the sweet twitter of wings, or sit on the tailgate under a darkening sky and watch the woodcock at their dusking flights and wonder where they'll be tomorrow.

# 8

# Different Drummer

At odd moments, my mind truant from more productive thought, I sometimes consider how my life would be different if certain things did not exist, think of what might be absent if it weren't for this or that. I don't mean to alarm any would-be writers, but this is what happens after nearly twenty years of making a living by sitting alone all day in a small room talking silently to sheets of paper or a word-processor screen. Like other socially acceptable forms of lunacy, it's actually rather pleasant.

The lack of some things—speedboats and ski resorts, opera, roller derby, or life in the big city—would not affect me in the slightest. For others, like trees and spring brooks, chamber music, the silky fur behind my dog's ear, and strong coffee before daylight, the difference would be profound.

And grouse. Were there no such thing as a ruffed grouse, about half the art in my house would disappear, my library would be almost barren of titles that begin with G, and neither my faithful old Blazer nor my favorite boots would have nearly so many miles on them. I would be bereft of some of the dearest friends I've ever known, the companionship of certain dogs, much of my sense of wild beauty, and nearly all the month of October. I would not have seen firsthand several thousand square miles of the loveliest, scruffiest country on earth, from Nova Scotia to northern Minnesota, nor spent priceless hours leg-weary, soaked to the skin, chilled to the bone, and warmed to the soul.

If it seems odd that a bird could so affect a life, I can only suggest that to a hunter and lover of wild things, a ruffed grouse is not simply a bird; it's roughly a pound and a half of pure cosmic energy extracted from the soul of wildness and modeled in the form and variety of nature itself. Nor is my life different in this regard from many others, for the ruffed grouse is the icon of a brotherhood whose traditions reach back two hundred years or more.

I have a notion ruffs were the first birds hunted for sport in North America. Certainly they were not the first birds hunted; turkey and geese probably own that distinction. But once New World civilization evolved to the point where the expenditure of a pinch of gunpowder and shot no longer had to be measured against maximum returns in the amount of food it produced—when, in other words, the colonial hunter could afford to indulge himself in a bit of sport—I suspect he turned first to the bird that was a familiar fixture in the eastern forests.

The ruffed grouse is only one of nearly a dozen species of grouse native to North America, but along with the now-extinct heath hen, it is the only one native to the eastern United States, and it occupies the greatest range of any—from Newfoundland to northern Georgia, all across New England and the Appalachians and the Upper Midwest, throughout southern Canada, from the Colorado Rockies to the Pacific Northwest and north to Alaska. In Colonial days it lived as far south as Alabama, occupying northern Arkansas, most of Missouri, eastern Kansas, and the Iowa woodlands in the Mississippi valley.

◆

According to skeletal fragments uncovered in several areas of the country, ruffed grouse have existed here for at least 25,000 years. And they have remained unique to this continent. In the 1670s Nicholas Denys twice attempted to transplant grouse from Nova Scotia to France. The birds survived most of the voyage well enough, but as the bemused Denys wrote, "when approaching France, they die, which has made me believe that our air must be contrary to their good." Others apparently took the explanation to heart, for I'm not aware of any subsequent effort to introduce ruffed grouse to Europe.

As settlers and early naturalists began taking note of the bird, no one seemed entirely certain what to call it. Those in Pennsylvania and the South preferred "pheasant" or "mountain pheasant." Northerners still prefer "partridge," which in New England typically comes out as "pa'tridge" and in the Upper Midwest often gets shortened to simply "pat." English naturalist George Edwards, who described the bird in his *Gleanings in Natural History*, published in London in 1758, called it "ruffed heathcock or grous."

Although a gallinaceous bird and therefore kin, a grouse is neither pheasant nor true partridge, and Edwards' choice ultimately came to be accepted as the official common name for the bird ornithologists know as *Bonasa umbellus*—though you'll sometimes hear it mis-colloquialized as "ruffled" grouse.

A ruffed grouse is so called because it wears tracts of broad, glossy-black, faintly iridescent feathers on either side of its neck. Both sexes have them, but the male's are larger and more prominent. In an amatory mood, or when putting on an intimidation display, the bird flexes specialized muscles that raise these feathers into a ruff more magnificent than any Elizabethan fop ever dreamed of owning.

As to the balance of their dress, I'll simply say that to my eye ruffed grouse wear the most beautiful plumage of any birds on earth. Every hue of the autumn woods except scarlet and bright yellow is in there somewhere, subtle, muted, and ineffably elegant.

Although basically divided according to two categories of color—gray phase and red or brown phase—their actual appearance covers an extraordinary range of variations. To some extent, the differences are broadly geographic; northern birds tend to be predominantly gray and southerners red, but there is much overlap. In fact, both gray- and red-phase chicks can hatch in the same brood.

Other differences in coloration show up in whole populations of grouse. No one has the ultimate explanation for this, but climate, particularly moisture, appears to be a key factor. Coloration benefits the birds by helping them blend into their environment, and populations tend to be relatively light or dark according to whether the local environment is light or dark. Grouse of the dense rain forests in the Pacific Northwest are darker colored than those of the drier country farther east. These regional differences have led taxonomists to classify ruffed grouse not simply as *Bonasa umbellus* but rather as a dozen subspecies. I won't inflict all the Latin and Greek on you, but by common name the races include eastern, St. Lawrence, Appalachian, midwestern, gray, hoary, Yukon, Vancouver, Olympic, Pacific, Idaho, and Columbian grouse.

Regardless of their racial and regional identities, ruffed grouse are ruffed grouse, and nothing sets them quite so distinctly apart from other grouse as their eerily beautiful displays of sound. From the moment European man first heard it, every naturalist has fastened upon drumming as the essence of the ruff's nature and mystique.

In several species of New World grouse—blue, sage, sharptail, and prairie chicken—the males have inflatable, brightly colored air sacs on their necks, which they use as an audio component to their elaborate, often dancelike courtship displays. Booming, as it's called, serves to notify females that males are present on the courting ground, an avian come-all-ye and invitation to a grand show.

◆

Most of the booming grouse are grassland birds and basically flock dwellers. Ruffed grouse are solitary woodland animals, and their environment amounts to a labyrinth through which the sexes have to communicate in order to find one another. Sound obviously is an efficient medium. Booming, with its rather airy tones, works well enough in the open grassland, but the ruff evolved a technique better suited to the woods. Moved by the springtime urge to seek tender company, a male grouse finds a spot that offers a good view of the surroundings and an escape route that usually leads downhill. A fallen log is the classic venue, but grouse will as readily drum atop rocks, earthen mounds, culverts, exposed tree roots, snowdrifts, or anything else that offers the right components. If it's a permanent structure, it probably has been used as a drumming site by previous generations.

Onstage, he struts around for a while as a warm-up, then anchors his feet, leans back against his tailfeathers, and fans the air with his wings. In beats at first distinct and then progressing to a blur, it sounds like a cranky chainsaw that sputters, hitches, races, and abruptly dies: *put...put...put...put...put...put..put...put. put.put.pututututututututut...put.* The whole sequence lasts six to eight seconds.

On a windless day the sound carries remarkably well, and it would be a virtual dinnerbell for every predator in the countryside but for the fact that at a frequency of forty cycles per second it's pitched well below the hearing threshold of even the great horned owl. You can feel it as much as hear it, this low-voiced throb that beats like the very heart of the woods. It has been likened to everything from distant thunder to "the kettledrum of nature's orchestra," all of which are apt enough. (English naturalist George Robert Gray, who in 1840 gave the ruffed grouse its generic name, apparently heard it with a somewhat different ear, however; *bonasa* is Latin for "wild bull.")

Predictably, early explanations of exactly how the bird produces its sound boiled with controversy and dissent. Some native peoples believed the grouse actually strikes the drumming log with its wings, and one tribe's name for it translates as "carpenter bird." George Edwards, relying heavily upon information supplied by American naturalist John Bartram, believed grouse drum by "clapping their wings against their sides." A hundred years later Audubon agreed, and so, as late as 1905, did Professor C. F. Hodge of Clark University.

Edwards quotes two other observers who offer different possibilities: Mr. Brooke of Maryland thought they simply beat the air, and La Hontan was convinced the sound came from the wingtips striking together. Frederick Vreeland in 1918 essentially agreed with La Hontan but believed the wings clap together behind the bird's back rather than in front.

Yet another man, clearly one with a sublimely subtle mind, suggested that ruffed grouse really do have air sacs, internal ones (which for some strange reason no one could find merely by dissecting a bird), and fanned its wings to pump air out through its mouth—in effect playing itself like a bagpipe.

Of these early observers, Mr. Brooke came nearest the truth. In 1874 William Brewster described the sound as the concussion of wingbeats upon the air. After many observations during the spring of 1921 Edmund Sawyer of the New York State School of Forestry drew the same conclusion and insisted that the sound occurs when the bird's wings are on the upstroke. In 1929, analyzing motion-picture film of drumming grouse, Dr. Arthur Allen of Cornell University effectively demonstrated that the wings touch only air but pointed out that the sound synchronizes with the downstroke.

Actually Sawyer and Allen were both right. Snapping downward, the bird's broad, incurving wings create a partial vacuum next to its body, and in the next instant as the wings reverse for the upstroke, air rushes in to fill it. Simply put, a drumming grouse produces a series of miniature sonic booms.

Though grouse are inclined to perform year-round, drumming peaks in spring and fall. Autumn drumming clearly represents statements of territorial imperative, proclamations by adult cocks that others are unwelcome on his turf. In spring it serves the dual purpose of warning off rival cocks while inviting visits by any and all hens in the neighborhood. Should another grouse show up, the drummer spreads his tail fan, perks up his ruff, and fluffs his feathers until he appears about twice his normal size—a display meant to intimidate another cock or send a hen into a romantic tizzy. Either way, the little guy can't lose.

And he'll know at a glance which is which, although to a human it's not quite so easy. Anyone who has examined a lot of grouse can make a good guess as to what sex it is just from the marking and coloration of a bird's breast feathers. The size of the ruff is another clue among adults but less obvious for juveniles.

The classic characteristic is the dark band near the tips of the tailfeathers. This typically is black, but it may also be silvery, a deep cordovan brown, an incredibly beautiful dark rose, or any of a dozen or more shades. On the whole a tail band of the same color on all feathers signifies a male, and if the band is interrupted in the two central feathers, the owner usually is a female. Similarly you might count the white spots on the rump feathers—one spot for a female, two or more for a male. Most males sport bright-orange eyebrow stripes under the feathers above their eyes. A combination of all these confers the best odds of being right, but the only infallible means of determining sex is to perform an autopsy.

The chap on the drumming log, however, is always right the first time. If the visitor is a hen, the cock performs in his grandest manner, swaggering this way and that like a dwarfish turkey, puffed and preened, the very image of a swashbuckling rake. If the hen is receptive she does a bit of girlish flirting in return, and they mate. Then her dashing consort dashes off to plot his next conquest, leaving her abandoned and in the family way.

Which is okay with her, since a steady diet of swashbuckling can be tiresome. Besides, she has business to get on with. Finding a suitable place somewhere nearby, usually next to a tree or stump or sizeable rock, she forms a little bowl in the litter of the forest floor and lays about a dozen eggs, one each day if the weather is fair, every second day if it isn't. Once the clutch is complete she broods it for just over three weeks.

Grouse chicks are precocious, out of the nest as soon as their downy fluff is dry, ready to assault the insect world with a vengeance. Although adults are largely vegetarians the chicks are voracious carnivores, packing away so much insect protein that they're able to fly within two weeks. By fourteen weeks they are virtual duplicates of their parents.

In preflight days they depend upon an ability to freeze motionless and upon their cryptic, mottled marking for safety—these and a mother who greets intruders as if she weighed fifty pounds. She can moreover put on a masterful act of fluttering off as though injured, tempting danger away from her brood. You'll find plenty of funny stories about cock grouse squaring off to defend their territory against everything from people to bulldozers, but they're no braver than a hen with a parcel of chicks.

The family group will stay together until fall, and then under what old-time grouse hunters call the Mad Moon the young disperse, seeking territories of their own. It will be a more or less difficult task depending upon how many birds make up the local population and how much good habitat the neighborhood has to offer. Where the habitat is limited they likely will find that all the best spots are taken, and some probably will be driven off by their own fathers, whose autumnal drumming is a sign that he can go on the warpath at a moment's notice.

Some of the young birds literally will be run ragged, ending up in suburban lawns, city parks, and other unlikely places. A portion will be snatched by predators. Others, dashing about in their desperation, will run afoul of powerlines, plate-glass windows, the sides of buildings, and other hazards. It's nature's way of culling the autumn flock, helping to ensure that the population stays in balance with the habitat and that only the strong and lucky are left to face the rigors of winter.

By the Mad Moon, the young birds have shifted to a vegetable diet, and the catalogue of what they're willing to eat is nearly endless. Over the course of a year, adult grouse consume the leaves, berries, nuts, seeds, buds, and catkins of everything from clover to hardwood trees. Fruits and berries of every kind seem to be a great delight—wild grapes, apples, gray dogwood, highbush cranberry, poison ivy, farkleberries, sumac, blackberries, rose hips, mountain ash, chokecherry, elders, mulberry, bittersweet, you name it; if it grows in grouse country and bears fruit, it's grouse food.

Catholic as their taste may be, however, grouse have evolved a special relationship with one plant in particular. It is no doubt an ancient affiliation, but only now, thanks to the work of the late Professor Gordon Gullion of the University of Minnesota, are we beginning to understand the nuances of how the ruffed grouse and the aspen tree interact.

To commercial timbermen the aspen is the sylvacultural equivalent of the cockroach—prolific, fast-growing, opportunistic, and ubiquitous across the Great Lakes region, where by no accident ruffed grouse populations are densest. No other single species offers so much to grouse. In the sprout and sapling stages the birds feed on the tender leaves and use the dense growth for cover. Older stands provide well-protected living and breeding habitat, and mature aspen offers a vital source of winter food.

Ruffed grouse are not migratory, and any animal that stays put through the long, harsh northern winter must be able to exploit its habitat with maximum efficiency. By fall grouse are well insulated, their outer feathers backed by soft, downy plumes called afterfeathers. Unlike most other birds, their legs are feathered nearly to the feet, and the feet themselves are prepared to deal with snow. Each fall a fringe of tiny appendages grows all round the edges of a grouse's toes. About an eighth-inch long, these firm little spikes effectively double the bearing surface of the feet, thus serving as snowshoes that help the birds stay atop the fluffy drifts.

Grouse even use the snow itself for survival. Snow is an excellent insulator, and grouse make their own miniature igloos by simply burrowing into a soft drift to spend a comfortable night or wait out a stormy day.

Having a warm coat, a set of snowshoes, and snug snow-cave are all helpful, but no matter how well equipped an animal may be it cannot survive without a supply of nutritious food that can be obtained with a minimal expenditure of energy. Any number of trees and shrubs provide buds and catkins, but none is of more widespread importance than aspen. The buds are rich in nutrients, and the twigs and branches are stiff enough that a grouse can perch comfortably at the ends, where the harvest grows, pack in a cropful of food in a quarter hour or less, and head back to denser cover where it's protected from predators and the elements.

Good as it is, though, the grouse-aspen relationship is not wholly benign. Researchers and hunters alike have long puzzled over the cycles in grouse populations. In six- to ten-year sequences, ruffed grouse alternately flourish and crash. Their numbers grow to astonishing levels and then suddenly, mysteriously, they disappear as if swept away *en masse* by some great unseen hand. In subsequent years, the populations rebuild, reach a peak, and crash all over again. For a long time no one could discover any prime cause, not predation nor weather nor hunting pressure nor disease. The best explanation seemed to be that the cycles are caused by a combination of circumstances and events too subtle to fully grasp.

As historical data accumulated, however, researchers began to notice that the cycles only occur in regions where aspen predominates. Further investigation proved that grouse make the least use of aspen during the periods when their numbers are declining or have bottomed out. Which is a thoroughly curious thing: If the aspen tree offers so much to the grouse, then smaller populations, with fewer birds competing for food, should make more use of it, not less. Could the aspen itself be the culprit?

Apparently it is. Buds are even more important to the tree than to grouse, and the more grouse that feed on them, the fewer buds the tree has left from which to sprout leaves and reproduce. Like some other species, aspen have their own means of defense. The loss of buds to a certain level triggers production of a chemical known as coniferyl benzoate, which suffuses the buds and makes them unpalatable. This, moreover, sets off a chain reaction from tree to tree within a community, so when one begins pumping up its quotient of coniferyl benzoate those nearby do, too. In regions where aspens are abundant, like the Upper Midwest, the phenomenon can spread over vast areas, abruptly knocking back grouse populations as a result.

For a graphic demonstration you have only to compare populations in the Upper Midwest with those in New England year by year. When birds are dreadfully hard to come by in Minnesota, Wisconsin, and Michigan, grouse hunters in Maine, where there's relatively little aspen, find them plentiful.

Left to themselves grouse and aspen can get along quite comfortably in their seesaw romance. If they couldn't, one or the other would have disappeared ages ago. Man, however, represents a wild card that has changed the game.

All across grouse range the early years of settlement often enhanced the forest habitat. Old woods are not grouse woods. The birds spend most of their time afoot, and mature forest with its dearth of low-growing vegetation offers neither protection nor a livelihood. Grouse prosper in young woods, in and around openings created by fire or wind or timber cutting, places where sunlight can reach down to promote a lush regrowth of sprouting trees and fruit-bearing shrubs, dense enough to screen against attack from the air, open enough at ground level to allow the birds room to move about.

In the beginning, settlement created exactly the right patchwork of small clearings, but as settlement evolved into civilization what was just enough soon became too much—too much land permanently cleared, too much abuse of remaining woodland by deliberate burning and free-ranging livestock, too much year-round hunting and snaring for subsistence and the wild-game markets.

What happened to ruffed grouse in my home state of Missouri is a synecdoche of what happened through much of the bird's eastern and southern range. Missouri never was as prime for grouse as the regions farther north, but in the 1840s the Ozark highlands were home to more than a half-million birds. Fifty years later the populations were breaking down under the weight of habitat destruction and overharvest. As was common throughout the country, the bulk of the harvest was not for sport but for profit. In 1880 ruffed grouse and prairie chickens brought three dollars per dozen in St. Louis, and in 1886 alone 19,728 ruffed grouse were sold in the game markets statewide.

The harvest was halted altogether in 1905, but since the impact of habitat destruction was, as always, a far greater threat, the closure did little to stem the decline. A statewide survey of resident game and furbearers conducted in 1937 estimated the entire remaining population at a hundred birds.

Efforts in the 1940s aimed at restoring grouse to Missouri came to naught, but in 1959 the Department of Conservation began a program of releasing birds, live-trapped in other states, into two areas where the habitat was particularly good and could be managed for the birds' benefit. By 1977, 799 grouse had been imported and released, and some populations had grown to the point where native birds could be trapped and relocated. By 1983 we had enough grouse that the hunting season was reopened in parts of four counties along the Missouri River, and other counties have reopened since. Although the program has not shown quite such spectacular success as the one that restored wild turkeys to Missouri—mainly because turkeys are less specific about the sort of habitat they need—the populations are healthy and should continue to grow.

Similar work in restoring grouse to their ancestral range is under way in several states—Maryland, Arkansas, Indiana,

Kansas, Tennessee, and Alaska. In many cases the programs are supported and conducted through cooperative efforts among state wildlife agencies, the U.S. Forest Service, the U.S. Fish & Wildlife Service, and The Ruffed Grouse Society, a private organization dedicated to research and management for the benefit of forest wildlife.

No animal can stand alone against the juggernaut of human self-interest. Fortunately the ruffed grouse is not without his advocates, and as with much American wildlife those most dedicated to the survival of grouse as a species are those who love him as a game bird.

There is none finer. Where they have little or no contact with man, grouse can be unbelievably naive, which led the colonists to nickname them "fool hens." You can find them so in remote parts of the West, birds so tame you can bop them with a stick, no more game than a barnyard chicken. In the northern and eastern woods, however, it's another story, and the classic ruffed grouse is wily, tough, and elusive, a master at confounding hunter and dog alike.

In my musings on what life might be without grouse one image recurs—of standing among a shower of aspen leaves swirling like a flung trove of gold coins on a sudden gasconade of wind, knowing there's a grouse nearby to lend shape and substance and a fiercely concentrated spark of wildness. It recalls feelings of peace and balance and the comfort of having a place in this lovely scheme where life and death intertwine to amplify one another. Some northern Indians heard the voices of the gods in the sound of a drumming grouse. The more I listen, the more I believe they were right.

◆

# 9

## November Song

November sings an old, sweet song in my memory, a melody laced with woodsmoke, early mist, and the hard smell of winter on the wind. It sings in brown and bronze, russet, crimson, gray, and faded green, sings with the texture of prairie stubble, sings in the colors of a cock pheasant, and with much the same character—by turns a swaggering dandy and a short-fused brawler, holding out idyllic promise and holding back a hard-bitten reservoir of strength, tough as barbwire, blunt as a cussword.

November and pheasants and my father all three twine indisseverably along the contours of memory. Dad was a bird hunter, fond of rib-sprung pointers and rattly pump guns, a man who found his deepest peace in the homey, gentle charm of bobwhite quail. But November Saturdays belonged to pheasants, as if neither could be complete without the presence of the other. The association remains with me still.

November is no cuddly pup of a month, not where I come from at least, and neither is John Ringneck. He is the roughneck scion of an enormous tribe, the order Galliformes, a group divided into six families, about 265 species, and hundreds of subspecies. Pheasants and quail together, 183 species of them, make up the family Phasianidae. More distant kinfolk include families of grouse, guans, megapodes, guinea fowl, and turkeys.

The Phasianidae, known variously as pheasants, partridge, quail, francolins, snowcocks, junglefowl, spurfowl, and myriad other names, are in human terms probably the most important family of birds in the world, since some of its members are the progenitors of domestic fowl. The subfamily Phasianinae, the pheasants themselves, is especially significant because it includes the jungle fowls of the genus *Gallus*, from which virtually all domestic chickens are descended.

Except for the Congo peacock of central Africa, the Phasianinae are native to Asia. The fossil record shows an early ancestor—a flightless bird that stood about five feet tall and weighed nearly three hundred pounds—living in southern Europe about 25 million years ago. In more recent times, however, pheasants and their nearest kin have occurred naturally only from the Caucasus east to Malaysia.

This isn't to say they resist transplantation, as dozens of species and subspecies now thrive over much of the world, carried hither and yon in mankind's age-old fascination with the pheasant. Many are aviary specimens; others, like the Lady Amherst and golden pheasants, are half-domesticated ornamentals whose job is to simply show off their sumptuous plumage. Still others are free-ranging, freewheeling, rough-and-tumble game birds.

They're a diverse lot, showing enough variations to drive a taxonomist to the bottle. The typical game pheasant's pedigree is such a mélange of crossbred bloodlines that its genetic background is almost impossible to fully unravel. This is partly because pheasants aren't all that choosy about whom they shack up with. Ringneck cocks have been known to breed with everything from barnyard chickens to turkeys (in the process producing mercifully sterile offspring). Man, too, has contributed to the pheasant's racial stew by transplanting various species into single populations with a fine disregard for purity.

The bird we now recognize as the "typical" pheasant is the product of two main genetic influences alloyed by a handful of tributary branches. Our American birds derived largely from the Chinese ringneck, seasoned with Korean, Manchurian, Japanese, and English blackneck strains. The Japanese green pheasant remains a distinct species, but all the rest are now known simply as *Phasianus colchicus,* with the thirty-one purer ancestral strains recognized as subspecies.

All this racial mixing actually has been to some advantage. The pure strains are tough enough, but over countless generations the smelter of hybridization and the crucible of natural selection have produced a magnificently resilient animal. He is a splendid mongrel, John Ringneck.

His arrival in the New World was not without false starts. He made his debut about 1733, when James Thompson, governor of New York Colony, stocked a dozen pair of English blacknecks on Nutten (now Governor's) Island in New York harbor. George Washington released several pair at Mount Vernon during his first term in office, and in 1790 Richard Bache, Benjamin Franklin's son-in-law, released others along the Delaware River in New Jersey. The Nutten Island birds did better than the rest and shortly colonized Nassau Island, as Long Island was called then, but none of the eastern populations lasted very long. The real beginning of John Ringneck's tenure in North America came a hundred years later and almost 3,000 miles to the west.

While serving as United States consul-general in China in the 1870s, Judge Owen Denny took a notion to send some Chinese ringnecks home to Oregon. Pheasants commonly were sold in Chinese markets, and Denny probably developed a taste for them. He knew nothing of zoogeography, but he did recognize a similarity in the climates of eastern China and Oregon's Willamette Valley, so he bought a few dozen live birds in Shanghai, stuffed them into wicker crates, and shipped them east under sail.

All of Denny's early exports came to bad ends before reaching North America (with the possible exception of one shipment the crew ate as a celebration when the vessel dropped anchor in San Francisco Bay). But in 1881 twenty-one birds in a shipment of thirty survived all the way to Seattle, and Judge Denny's brother released them on the family farm near Peterson's Butte, west of Corvallis.

The birds prospered as newly stocked pheasants usually do when the habitat is right. In ten years they represented a population that may have been as dense as one bird to each five acres

in the Willamette Valley; this is an area 40 miles wide and 180 miles long, and if my arithmetic is right, that amounts to 921,000 birds. Oregon hunters reportedly celebrated a seventy-five-day bird season in 1892 by bagging 50,000 pheasants.

◆

New pheasant populations typically follow a boom-and-bust cycle in adjusting to local habitat, and Oregon's bonanza crashed and then stabilized in the mid-1890s. John Ringneck unquestionably had come to stay.

Stocking efforts all across the country came in fits and starts over the next thirty years. Those in the East and West fared well enough to establish adequate populations, but it wasn't until the birds reached the midwestern prairies that pheasants found their real North American home.

The land once supported a rich biome of grass and prairie forbs, but by the time John Ringneck came along it was largely corn country stretching from Pennsylvania to the Dakotas in a wide swath, covering all or parts of Ohio, Indiana, Michigan, Wisconsin, Illinois, Iowa, Minnesota, Kansas, and Nebraska. He arrived at the changing of an era, just as the tallgrass and the prairie chicken slid into history along the steel moldboard of the breaking plow, as Indian grass, bluestem, and ripgut gave way to orderly rows of cash grain. The pheasant filled the niche as if nature had it planned that way all along.

Pheasants are not wilderness birds anywhere. Seed-eating John Ringneck thrives instead on the bounty of row-crop farming, in the manmade environment that prairie chickens cannot tolerate and at latitudes too brutally cold for bobwhite quail. Besides, the country suits the ringneck's personality—wild, exuberant space under a windy blue bowl of sky, the climate in a continuous shift between blistering summer heat and winters cruel as iron.

Simply to survive in open country wracked by such atmospheric extremes demands a rugged body and a durable spirit; to prosper there requires special gifts. Whether dressed in the cockbird's gaudy Jacob's coat or the hen's more subtle buff and brown, a pheasant is a splendid physical specimen. At an average weight of just under three pounds, a mature cock is a bruiser by upland-bird standards. Hens average perhaps a pound less, but both sexes carry enough body mass to withstand hard knocks and deadly cold. Holed up in some sheltered spot, pheasants can easily wait out a three- or four-day blizzard without food, and some have been known to survive a solid month without eating at all.

Pheasants also are highly resistant to disease and internal parasites, perhaps more so than almost any other North American animal. Their fondness for dustbaths helps keep them similarly free of mites and other exterior vermin.

All of the Phasianidae are ground-dwelling birds, and a terrestrial life is fraught with perils. As one means of defense pheasants and their kin have fairly short wings shaped for rapid takeoff, a characteristic probably most efficient in quail and grouse. Because a pheasant's ratio of wing area to body weight is lower, takeoff requires more effort, but John Ringneck is no slouch in the air. After the initial flush, accomplished at about three wingbeats per second, pheasants level off into a pattern of alternate wingbeating and gliding that can top forty miles per hour.

Even so, a pheasant's main defensive tools are his long, remarkably muscular legs, which are reinforced at the thighs by sesamoid bones. These thin, splinterlike bones serve both to support the ligaments and to enhance the angle between the leg and its muscles. He can glide through thick vegetation as smoothly as a cat if he needs to, skulking along with head and tail carried low. But given some real running room—in short grass or between corn rows—John Ringneck raises his head, cocks his tail at a high angle, and pops the clutch. He has a stride of about eighteen inches at full stretch, and even the fittest dog has to work to run him down. In a footrace with a pheasant a man hasn't a chance unless he's a trained athlete.

As if sheer speed afoot weren't enough, the bird usually has the advantage of a headstart. You can sooner catch a wolf by surprise. A pheasant's cold, bright-yellow eyes don't miss much, but hearing is by far its keenest sense, considerably more acute than a mallard's, perhaps even more sensitive than an owl's. Many animals can detect approaching thunderstorms long before they're audible to humans, but no one fully appreciated the extraordinary acuity of pheasants' hearing until World War I, when penned birds in England unmistakably reacted to the Battle of Dogger Bank, which was being fought 216 miles away. Observers later documented the birds' ability to detect cannonfire at a distance of 320 miles.

Like some others among his clansmen, a cock pheasant is armed with spurs on his lower legs. In a first-year bird they're scarcely more than rough, cone-shaped bumps, all but useless as weapons. If he survives to see it, however, John Ringneck enters his second autumn sporting a pair of inch-long daggers with needle tips, and by then he's learned how to use them. Every old pheasant dog—and a lot of pheasant hunters—can show you the scars to prove it. I have some myself, courtesy of a wounded cock I once picked up by the neck.

A pheasant no doubt takes some satisfaction in being able to stick dogs and careless hunters, but his spurs have an even more serious use as part of his love life. Like most of the larger, more ornate species of Phasianidae, cock pheasants are both polygamous and territorial during the mating season, staking out domains in which to woo their lady friends and defending them with beak, spur, and wing against any other cock who gets too close. He's willing to fight to the death and sometimes does.

These are busy, critical weeks from the tag-end of winter until mating activity peaks in April. When he's not slugging it out with his colleagues, he's strutting about in full glory, crowing and fanning his wings to attract hens, generally showing the ladies what a lordly specimen of pheasanthood he is. He's capable of breeding with as many as fifty hens, although population densities rarely provide the opportunity. His harem more likely will number from six to ten.

Which is enough to keep him occupied, because he courts each one in an elaborate ritual, parading in circles with his plumage ruffled and his tail partially fanned, hissing and bobbing like a swashbuckling rake. If it's too early in the season the hens may pay only token attention or may ignore him altogether. In that case the poor fellow has to go through the whole production all over again, day after day until the females' hormones catch up with his. Force is out of the question; for all their hair-trigger belligerence, cock pheasants assemble and maintain their harems solely by persuasion.

By the time she gets her nesting act fully together and starts brooding her clutch of about a dozen eggs, a hen becomes a serious parent—which is both boon and bane. On the one hand she and the cock can put up a good defense against predators. Nest predation is heaviest early in the season when cover is still relatively thin. It is often the work of skunks, ground squirrels, crows, and snakes.

On the other hand, once springtime vegetation grows dense enough to conceal the nests from marauders a new and far deadlier threat appears in the form of mowing machines. In some fields the loss may be as high as 40 percent, and the damage isn't only a matter of smashed eggs. A hen will quickly renest if her clutch is destroyed early—but not if she herself is butchered in the blades of a sicklebar. Her devotion to the nest is so steadfast that she often refuses to leave it until too late. The documented loss of hens to mowing machines has in some areas reached as high as 30 percent of the nesting population. Landowners can promote pheasant conservation simply by waiting to cut hay until after the hatch.

New life appears in a pheasant nest about twenty-three days after incubation begins. Since she doesn't start incubating until the clutch is complete, some of the eggs may be nearly two weeks older than others, but all of them hatch within a twenty-four-hour period. Newly sprung from confinement, a pheasant chick weighs less than an ounce, but it arrives fully dressed in down and begins packing away protein-rich insects as soon as its feathers are dry.

Rain and chill are primary threats to their early survival, and the hen continues to brood her chicks for a week or more after they hatch. Very young chicks apparently give off little scent, and immobility is their main defense. They also use a distress call, and the hen responds with the gritty intolerance of a mother bear. Depending upon the nature of the danger she may distract an intruder with a classic broken-wing routine or wade in like a feathered swarm of bees. Two pounds of angry hen pheasant is often more than enough to convince even a dog-sized predator that there's a better meal somewhere else. For their part the chicks develop quickly. Vaned feathers begin to replace the natal down within a few days, and they can fly by the time they're two weeks old.

Nonetheless nature has no intention that every animal should survive, and an average of three chicks in a typical pheasant brood won't live to the age of seven weeks, victims of weather, predation, or mischance.

As with most small animals, youth and age are relative terms. In captivity pheasants can live as long as eight years; average lifespan in the wild is about nine months. Hens live a bit longer as a rule, in part because they're physically more resistant to weather and starvation, and because they are not legal game during the hunting season. Game biologists realized early on that given proper habitat hens are the key component in maintaining high pheasant populations. Thanks to their polygamous sexual habits, only one of every ten cocks needs to survive through to the breeding season, as compared with a 55 percent survival rate necessary for hens. Where annual populations are composed almost entirely of stocked birds—as in England, Europe, and on American shooting preserves—both sexes are fair game, and some states have allowed a limited harvest of hens. But among most wild, self-sustaining populations dense enough to warrant hunting, a hen in a game bag is a sure ticket to an expensive visit with the local judge. The 90 percent surplus among cocks actually is more theoretical than real, since the harvest is never that high over a large enough area to impair the next year's breeding stock.

Land use, on the other hand, is critical, and the bulldozer and plow are far deadlier than the gun. Since pheasants and agriculture are inseparable, populations swell and ebb according to how much or how little land is left fit to see the birds safely through the winter and give them a place to nest come spring. Pheasants are hardy animals, but not even they can survive the harshness of a northern prairie winter without some hard cover for shelter and some waste grain and annual weeds for food. Pheasants, especially hens, flock up in dense cover during the winter, flocks that may number from a hundred to as many as a thousand birds. That many in one place is a spectacle. Finding winter-killed birds whose lives were traded for a few extra bushels of grain prompts an altogether different feeling—bad enough if you're simply an admirer of John Ringneck's beauty and brass, sadder still when it happens at home.

When I was a lad in southern Iowa there were no pheasants in any of the two lower tiers of counties. It wasn't that they couldn't live there but rather that they didn't need to; the central and northern counties offered all the habitat Iowa's millions of pheasants could want. Since the late 1960s, however, Iowa birds have drifted steadily southward, moving not only into the southern counties but well into northern Missouri. They're now abundant on the same farms where I once could hunt quail all day and never see a pheasant—but I can also drive all day through northern Iowa country where my father and I found thousands of them and see nothing but bare fields, plowed within inches of the roads and rolling for miles, unbroken by the brushy swales that were prime bad-weather habitat. Pheasants there are all but a memory.

I'm not unhappy to see John Ringneck moving south to meet me, but he can't keep running forever. Nature itself has drawn the line, roughly from Baltimore to San Francisco, and by far the majority of North American pheasants live north of it. There are some isolated populations much farther south, but for the most part John Ringneck steadfastly refuses to live and breed in great numbers much below the fortieth parallel.

Exactly why is anyone's guess, although the bulk of evidence so far points to a combination of factors in which climate and minerals play important roles. At latitudes where ground temperatures reach eighty to ninety degrees Fahrenheit during the nesting season in May and June, pheasant eggs show a sharp drop in the hatch rate. But temperature alone isn't the whole story. Pheasants from breeding populations in hot, dry southern California have shown little reproductive success when transplanted into the Southeast, so humidity may be a factor, too.

The mineral theory, first advanced by Aldo Leopold about sixty years ago, argues that calcium carbonate is the key. Professor Leopold pointed out that midwestern pheasant range almost exactly duplicates the extent of glaciation that scoured North America until about 10,000 years ago—a region rich in ground-up limestone. Pheasants need about twice as much calcium as they are able to get from normal feeding, and they make up the difference with limestone grit that has a high content of calcium carbonate. During the breeding season the need for calcium is absolute. If it's not available reproductive success declines; eggshells are too fragile and the embryonic chicks' bones don't develop properly.

Not just any limestone will do, however. In many parts of the South available limestone is dolomitic, with a magnesium content sufficient to chemically prevent the calcium carbonate from being absorbed into the birds' bodies. The proper minerals therefore must be available physiologically as well as physically.

Neither quail, grouse, nor turkeys are as calcium-sensitive as pheasants, and the problem may be that as an introduced animal the pheasant has not evolved in North American habitat to the same level of mineral requirements as native birds. In any event the whole matter could become the cornerstone of disaster if habitat destruction continues, squeezing John Ringneck to death against an invisible barrier.

Should that happen I can only hope I'm not alive to see it, because he's too old a friend for me to be comfortable without him, too much a part of the boy I used to be. I'd miss hearing the raucous, rusty-gate clamor of his springtime crowing, miss seeing his high-tailed sprint through the burnished gold of a cut cornfield, miss the angry cackle of his flush, his wily ways and go-to-hell attitude.

And I'd miss trudging carward with my dog at the end of the day, a couple of fat roosters making a comfortable burden in my game pocket. At those times my father is always with me and November always sings its old, sweet song.

◆

# 10

# Symbol of History, Symbol of Hope

Although Benjamin Franklin made no objection when the Congress of the Confederation chose the bald eagle as the emblem of the United States, he later confessed that he thought the carrion-eating eagle to be "a bird of bad moral character." In a 1784 letter to his daughter he wrote, "The turkey is a much more respectable bird."

While Franklin never actually proposed the wild turkey as the federal emblem, it wouldn't have been a bad idea if he had. Wild turkeys and bald eagles are both unique to North America, but while the snowy-headed eagle only *looks* regal and proud and majestic and all the other drippy adjectives we hyperbolize upon certain animals, the wild turkey is almost exactly what it appears to be—powerful, handsome, intelligent, circumspect, tough-spirited, and highly adaptable. In search of a symbol, a country could do worse.

Besides, the new nation owed the turkey some debt—if not adoption as totem then at least a measure of thanks for having fed several generations of colonists who found that it offered excellent return on the investment of a rifle ball and a pinch of powder. In his book *New English Canaan*, published in 1637, Thomas Morton puts it this way: "Turkies there are, which divers times in great flocks have sallied by our doores; and then a gunne (being commonly in a redinesse), salutes them with such a courtesie, as makes them take a turne in the Cooke roome. They daunce by the doore so well. Of these there hath bin killed, that have weighed forty eight pound a peece. They are many degrees sweeter than the tame Turkies of England, feede them how you can."

Unlike some New World creatures, turkeys were recognized instantly by the early colonists. The "tame Turkies" Morton mentions were descendants of wild birds that the Spanish took from Mexico to Europe in 1519. These domesticated birds reached England in 1524 and were promptly given the name *turkey*, which was the same name the British applied to guinea fowl imported from Turkey. A hundred years later the Pilgrim settlers brought a few birds with them to the New World. Which was a prudent move but unnecessary; at the time most of North America teemed with wild turkeys from southern Maine to the tip of Florida, from the northern Colorado Rockies through the Southwest and on to Central America.

The founders of Plymouth Colony didn't know that, of course, nor did they much care beyond the fact that turkeys offered good eating and plenty of it. Only a few might have been interested to know that the birds they brought shared the same ancestral genes as the wild ones over which they muttered mealtime prayers.

Nearly the same, anyway. While we now know the "typical" wild turkey as *Meleagris gallopavo*, differences in coloration and size create five distinctive races. The Pilgrims' bird, *Meleagris gallopavo silvestris*, the eastern wild turkey, is the most widespread.

◆

Other subspecies include the Florida turkey, which occupies most of the Florida peninsula; Merriam's turkey, found through much of the West and Southwest; the Rio Grande turkey of southwestern Texas and northeastern Mexico; and Gould's turkey of Mexico and the southwestern United States.

A sixth race, *M. g. gallopavo,* or South Mexican turkey, is now believed extinct. Domesticated ages ago by the Mexican Indians, the South Mexican bird contributed the original broodstock for the turkeys the Europeans knew.

Only one other species of wild turkey exists anywhere in the world: *Agriocharis ocellata,* or ocellated turkey, so named because its plumage bears eyelike markings similar to a peacock's. It once lived throughout the Yucatan Peninsula, west to the Pacific and south through Guatemala. Now, diminishing populations survive in parts of the Yucatan, in northern Guatemala and northern Honduras, holdouts against human encroachment.

The other species once suffered similar straits. Imprudent land use and unregulated harvest combined to so ravage the population of eastern turkeys that Pennsylvania, West Virginia, Kentucky, Missouri, Arkansas, and Louisiana now hold the only truly indigenous stock. Which is not to say these are the only places where turkeys live nowadays. On the contrary, the wild turkey now thrives over much of its former range, thanks to the successes of modern wildlife management. And in the process of restoration we've learned a great deal about the biology and private life of a truly splendid bird.

Turkeys are gallinaceous fowl, kin to our native grouse and quail, and are by far the largest members of the family. The different subspecies vary in size somewhat, and the weight of each individual varies according to season, but they're all big birds—though not as big as Thomas Morton led his readers to believe. Hens average about half the weight of males, and mature toms of the eastern race average about eighteen pounds. Still, birds of twenty-two to twenty-five pounds are not uncommon, and occasionally some hunter will bag one that runs nearly thirty pounds. Even so, whatever scales showed Morton's birds at forty-eight pounds clearly were in need of repair.

Turkeys are among the world's heaviest birds that are capable of flight, and for all their size they're astonishingly quick and powerful in the air. With the relatively short, broad wings typical of the Galliformes they're built for quick takeoff, and they do so with great panache.

When my Brittany girl, October, was about six months old she pointed what we both assumed was a covey of bobwhites in a patch of brush no larger than a toolshed. When six turkeys came thundering out literally under our noses my heartbeat took instant, and seemingly protracted, time off the job. I don't know exactly what Tober thought about those feathered things twice her size, but instead of chasing merrily after them as she would have with quail she simply sat back on her haunches and watched them go, her eyes round as golfballs and her expression the nearest thing to awestruck that a puppy face can form. She's pointed single turkeys at least once every hunting season since, and even now as a *grande dame* of twelve her reaction is still the same.

So is mine, for that matter. I don't know how many hundreds of turkeys I've seen on the wing, but it's a perennially startling affair. Part of it is sheer size but another part is that you're far more likely to see turkeys on the ground. They like to roost in trees (and flushing them from overhead in the stillness of sunrise woods when you don't know they're there in the first place is another exercise in cardiac arrest), but they spend their days on the ground, scratching in the leaf litter for acorns, pecans, beechnuts, and other mast, or foraging for grasshoppers in a rank summertime meadow.

They prefer to protect themselves by using their legs as well. At the first hint of danger a turkey ducks its head and scuttles off, following an instinct that leads to the densest undergrowth around. I don't know if anyone has ever clocked one, but a turkey going flat out can easily outpace a man and even give a dog a run for its money. Surprise one at close range and it might flush; more often it simply fades away in a quick patter of feet.

Surprising a wild turkey under any circumstances takes some doing. It's equipped with extraordinarily sensitive eyes and ears, nearly able, as one old turkey hunter put it, to see through a thin layer of rock and to hear you thinking. Ramble the woods in turkey country and you can find signs aplenty—tracks, scratchings, droppings, and the occasional castoff feather—but if you see the birds themselves it's usually only a glimpse as they disappear in the distance.

But if you're willing to work up some virtuosity with a turkey call, to be in the springtime woods before dawn, and to practice the arts of stealth and camouflage, you might get to see the finest show in turkeydom.

Springtime is mating season. The adult toms have gone on a feeding jag in preparation and are the heaviest they'll be at any time of year, owing to a thick layer of gelatinous, fatty material that has formed on their breasts. They'll draw upon this breast sponge, as turkey biologists call it, for energy during the breeding period.

And they'll need all the energy they can get, because like most gallinaceous birds turkeys are polygamous. Each dominant tom assembles a harem of several hens and maintains his little coterie as best he can by a strutting display in which he ruffles his feathers, draws in his neck, lowers his wings, fans his tail, and generally turns himself into a wonderfully imposing figure roughly the size of a Subaru. His featherless head, neck, and wattles change color according to his humor—blue, white, and, especially when he's in an amatory mood, red. (I sometimes wonder if this color scheme had anything to do with Ben Franklin's interest in the turkey as a national symbol.)

To make himself more imposing yet he often distends the fleshy growth on his forehead—called a caruncle, dewbill, or snood—so that it hangs down alongside his head. His beard, a curious tuft of bristly feathers growing out of his breast at the base of his neck, sways as he struts. (Although the beard is basically a male characteristic, some hens also have them and a few males have more than one. No one knows exactly what function it serves but among toms there is some loose correlation between the length of the beard and the age of the bird.)

The display has a component of sound as well—a rattle and click of feathers and a soft, eerie thrumming like the vibrating twang of a huge rubber band.

Most field guides will tell you that male turkeys gobble and females talk in a note that sounds like *quit!* Actually their repertoire is much more extensive than that. Toms do gobble, especially in the spring, to advertise their presence, call their harems together, and define their territories. They often gobble from their roosts just at dawn and for some reason frequently gobble in response to a clap of thunder, a crow calling, or an owl hooting. Imitating an owl at the right time on a spring day is a good way of learning whether turkeys are nearby.

Both toms and hens make a wide range of sounds, although some are more characteristic of one sex or the other—putts, clucks, whines, kelps, cackles, purrs, and an alarm call that sounds like *prttt!* Ask a turkey hunter about the lore of calling and you'll hear all these and more, including the tree call, the lost call, the fly-down cackle, and a nifty little figure known as the "kee-kee run." Talking good turkey requires some practice, but if you can tempt a big tom into a display, either as the prelude to harvesting a holiday dinner or just to watch the show, it's well worth the trouble.

The ocellated turkey's courtship performance involves a similar routine although its gobble reportedly is quite different. *The New Dictionary of Birds* says it sounds like *ting-ting-co-on-cot-zitl-glung.* Honest. I suppose it's possible that the ocellated turkey speaks Mayan or the ancient Nahuatl language of the Aztecs, but if you can make *ting-ting-co-on-cot-zitl-glung* sound anything like a bird, I'd like to hear it. I tried, but my friends made me quit.

Regardless of the sweet-talk involved, what happens after the breeding season is much the same among all turkeys. The hens scratch out a rough nest, usually under a bush of some sort, and lay their eggs. Eleven is an average clutch although more than one hen sometimes will use the same nest. The eggs hatch after twenty-eight days' incubation, and the chicks immediately leave the nest to start feeding on insects, seeds, fruits, greens, or anything else that's handy. Like all young things they are vulnerable to predation, disease, and the vagaries of weather, particularly cold spring rains that may chill them fatally. Within a year as many as six of every ten likely will have died of one cause or another.

A 60 percent mortality rate may seem high, but by nature's standards it really isn't. The mortality rate of itself doesn't lessen any species' ability to prosper when given suitable habitat and a measure of protection during critical times of the year. But these, in their absence, nearly brought the wild turkey to doom.

Turkeys are forest-living birds, and from the moment settlement of North America began their habitat started to disappear. What timberland wasn't cleared for farming or lumber or fuel often was given over to free-roaming livestock—all of which deprived turkeys and other woodland wildlife of the food and shelter they needed to survive. In many areas such as the Ozark Mountains of Missouri and Arkansas settlers practiced a traditional, thoroughly misguided ritual of burning the woods every spring, usually just at turkey nesting time.

Indiscriminate harvest also took its toll. Like other game, turkeys were shot and trapped year-round. Where they existed at all, game laws prescribing hunting seasons and bag limits were largely unenforceable and almost universally ignored. As the frontier expanded, wildlife dwindled in its wake. By the 1930s dozens of species were gone, nearly or altogether, from vast areas of their former haunts all across the United States. The nationwide population of wild turkeys, once perhaps as high as 10 million, stood at about 300,000 in 1950.

What happened next, as the art and science of wildlife conservation grew from infancy into full bloom, is something of a marvel. The white-tailed deer is one of conservation's great success stories. The wild turkey is another, and nowhere is that story more poignantly clear than in Missouri.

By the best estimate upwards of a quarter-million turkeys ranged the woodlands of presettlement Missouri. In 1940 about 4,000 remained, comprising small remnant flocks scattered through the rugged Ozark hills. Deer-restoration projects were coming along nicely, but turkey projects were not. Biologists had released as many as 500 wild-domestic hybrid birds every year since 1925, only to see them disappear with scarcely a trace. A ban on turkey hunting imposed in 1938 similarly showed no appreciable effect as populations continued their decline.

Finally in 1940 the Missouri Conservation Commission charged a three-member team of biologists headed by A. Starker Leopold, son of conservation guru Aldo Leopold, with the task of finding some solid answers.

Guided by principles inherited from his father, Leopold saw habitat management as the key to wildlife management: Determine an animal's specific needs, provide habitat that meets those needs, and the population will respond. Today this seems self-evident. Fifty-five years ago it represented the dawn of a new age.

For five years Leopold and his colleagues studied the wild turkey flock on the Caney Mountain Turkey Refuge in Ozark County, on the Arkansas border, manipulating the habitat components of food, water, and cover in an effort to find the optimum combination of type and proximity. Hen turkeys, they noticed, prefer to nest within three hundred feet of water, and by building a series of ponds at likely locations they were able to significantly expand nesting habitat within the refuge. Food plots sowed with plants that turkeys favor and located near escape cover showed similar good effect.

By the end of World War II the biologists had developed the framework for a long-term program of restoring turkey habitat throughout Missouri and repopulating it with released breeding stock. As Leopold's team worked at Caney Mountain, studies at other Missouri refuges conclusively demonstrated that hybrid turkeys are poor candidates. Although easy enough to rear on game farms, hybrids are ill-equipped to survive anywhere else. They lack sufficient wariness to avoid predators and are inclined to socialize with fully domestic turkeys, which exposes them to all the diseases that plague domestic fowl. In 1943, realizing that its release efforts had actually amounted to an eighteen-year coyote-feeding program, the Conservation Commission quit fooling with game-farm turkeys altogether.

To create a source of suitable birds the Commission established more refuges in the Ozarks, managed them intensively for high turkey production, and in 1946 began trapping and transplanting. Inevitably there were some hitches, particularly during the dreadful droughts of the early 1950s when the turkey population statewide plunged to fewer than 2,500. By 1957, however, conditions were improved and from there the Missouri turkey-restoration program, headed by biologist John Lewis, went into full swing.

Over the course of the project, which was officially declared complete in the spring of 1979, 2,611 turkeys were trapped and released at 142 sites in 87 counties. A dozen to two dozen birds typically were released at each site in a ratio of two hens for each tom. Sites were carefully selected to provide the best available habitat, and through cooperative agreements local landowners helped protect the birds and improve their living conditions.

And the turkeys responded. Following a virtual textbook scenario they simply went forth, were fruitful and multiplied, colonizing new territory at a breakneck pace. Their numbers were such by 1960 that a hunting season was reopened. In 1978, using a production index and census figures based on years of accumulated data, John Lewis believed the statewide population to be roughly 250,000 birds—about the same as it was before the white man arrived. In an extraordinary contrast to the situation in 1940, when turkeys were known to exist in only 13 Missouri counties, all 114 counties in the state held huntable populations by 1985. Now the population supports a statewide legal harvest of about 50,000 turkeys each year, and their numbers continue to grow wherever suitable habitat exists.

The variety of what turkeys consider suitable living space has come as some surprise. As biologists continued to study the birds and to refine management techniques they learned among other things that the wild turkey is a remarkably adaptable bird. Though not as tolerant of human presence as deer, turkeys have taken to agriculture extremely well. In fact the densest populations in Missouri now occur in the northern counties, which also hold most of the state's farmland. Woodland, once considered the sine qua non of turkey habitat, makes up less than 20 percent of some of those counties.

Consequently we now understand that "turkey habitat" isn't so simple after all. Actually turkeys require three distinct habitats—one for nesting, one for the summer and autumn, and one for winter. Nesting habitat is the most varied, but it must contain thick vegetation at ground level and have a permanent source of water nearby. During the summer and fall hens and poults thrive best in hayfields, meadows, and open woods where seeds and insects are plentiful. In winter all turkeys need hardwood forests where at least half the trees are mature and therefore mast-bearing. Missouri turkeys feed on acorns year-round to some extent, but acorns account for more than a third of their diet during late fall and winter, and such high-energy food is a crucial factor in winter survival. In years when the mast crop fails turkey populations decline.

And then they recover, sometimes almost as if by magic. The real reason, of course, is that the wild turkey is an enormously resilient animal, and their tough spirit is standing them in good stead over much of their former range. As the restoration project flourished in Missouri other states initiated similar efforts, many of them using birds trapped from the burgeoning Missouri flock. Now, turkey-restoration programs exist all over the country, and huntable populations live in all of the lower forty-eight states and in Hawaii. Only Alaska has no turkeys.

The National Wild Turkey Federation—which since 1973 has supported research and habitat restoration, encouraged public support, and fostered greater awareness of wild turkeys on a national scale—estimates the current population in the United States at 4 million and growing. Turkeys now occur in ten states where they did not exist when Europeans first arrived in the New World, and they occupy more square miles of habitat than any other North American game bird.

Even at that we are still short of the saturation point. About 50 million acres of suitable habitat remain unpopulated, and a cooperative effort now underway among state and federal agencies and the Wild Turkey Federation—a project called Target 2000—intends to have those areas stocked by the end of this century.

I had never seen a wild turkey until I moved to Missouri twenty-five years ago, and I must confess I'm not an enthusiastic turkey hunter, mainly because I can't sit still long enough to do it properly. But it has been my good fortune to have been involved in the turkey renaissance during its most exciting years and to have known most of the principal biologists and game managers as colleagues and friends. I think of these men whenever I hear the turkeys talking among themselves on the ridges around my farmhouse and when I watch them on summer evenings just at dusk as they come down into the hollow and forage along the meadow at the edge of the woods.

At such times it's clear that the wild turkey has indeed become a symbol, as much so as if our country's founders had agreed with Ben Franklin, a symbol whose significance we may not yet fully appreciate. The dedication and perseverance that my old friends and dozens like them have shown seem to me good cause for hope—not only for the wild turkey but for other wildlife as well. And not the least for ourselves.

# 11

# The Voice of the Turtle

I don't suppose I was any more literal-minded than the average kid, but I spent quite a few years in a state of profound puzzlement after I first heard someone read the twelfth chapter of The Song of Solomon: "The flowers appear on the earth; the time of the singing of birds is come, and the voice of the turtle is heard in our land."

Flowers and birds made sense, but no matter how much I listened to the little sliders and painted turtles from the five-and-dime and the box turtles I found trudging the woods, not one ever uttered a word. Neither did the big sliders and the soft-shells that lived in the rivers and ponds Dad and I fished. Only the snapping turtles we came across now and then made any sound at all, and I couldn't believe anybody would associate blooming flowers and singing birds with the menacing, gaspy hiss of a snapper.

Actually though, I was hearing the voice of the turtle all summer long—a clear, soft ooh-OO-hoo-hoo that seemed to reverberate from all directions at once, especially on sultry, overcast days. My mother said it was the voice of the rain crow. When you're six or seven you don't tend to be skeptical of what your mother tells you, so it was quite a few years before I managed to sort out the facts—namely that the bird a lot of midwestern country people call a rain crow is actually a yellow- or black-billed cuckoo; that the song they ascribe to the rain crow actually comes from the mourning dove; and that the turtle the Solomon writer had in mind isn't literally a turtle but rather a dove as well. It was quite some time later when I learned that a turtledove is, strictly speaking, an Old World bird and that the name derives from the Latin word for dove, *turtur*. But by then I'd already twigged to the fact that you can't always take words—or birds—at face value.

This certainly is true of the mourning dove. Watch one puttering about on the ground or perched dozing on a power line and you'd never guess it's one of the most abundant and widely distributed birds in North America; that it nests in every one of the forty-nine continental states, most of Mexico, and much of South America; that it can maintain a cruising speed of about forty miles per hour, hit sixty in an instant if it wants to, and cut some incredible maneuvers in the process; that it's an exceptionally fine game bird; or that it's the only North American game bird truly able to prosper living in close proximity to man. Nor from a distance can you see that its drab-looking plumage is really a subtle combination of gray, brown, buff, blue, pink, olive, black, white, and even a bit of iridescence.

What you would notice at almost any range is that a dove is some sort of pigeon. The small head, the short, rounded bill and short legs, the profile, the stance—everything is distinctively pigeonlike.

Which is only to be expected, since doves and pigeons are all part of the same family, the Columbidae, and the only real differences among all 289 members worldwide are size, location, plumage, and semantics. Technically, a dove is a pigeon is a dove. Their only close relatives are the dodo and the solitaire, both of which were extinct by 1800, and the African sand grouse. The mourning dove is first cousin to the passenger pigeon, which went extinct early in this century—and is similar enough in fact that mourners often were mistaken for passengers for some years after *Ectopistes migratorius* disappeared.

Most of the Columbidae live in the tropics, and most of the rest live someplace other than here. Argentina, for instance, is home to twenty-three species of pigeons and doves, including *Zenaida maculosa*, the eared dove, which is the keystone of Argentine bird shooting, so numerous and so destructive of agricultural grain that the government used to pay a bounty of one shotgun cartridge for each pair of dove feet delivered to the proper authority.

Even though we can't match such places species for species, North American fauna is not entirely dove- and pigeon-poor. Some, such as the spotted dove, ringed turtledove, and rock dove or domestic pigeon, were brought here from other continents. As natives we have the red-billed, white-crowned, and band-tailed pigeons, the ruddy and common ground doves, and the Inca, white-tipped, white-winged, and mourning doves.

The band-tail occupies a fair amount of country in the West and the common ground dove somewhat less in the South and Southeast. All the rest are either highly restricted in their territories or venture into the United States only along the southwestern border—except the domestic pigeon and the mourning dove. The rock dove's North American range may be a bit more extensive than the mourner's but not by much, and while the common pigeon is a noisome pest in most urban areas, the mourning dove is a charming, unobtrusive bird even where it appears in great numbers during the fall migration.

Among our native species, the mourner and the white-wing are remarkably similar. The white-wing, which is plentiful in parts of Mexico and along the border, is slightly larger, has a shorter, squared-off tail, and bears broad white patches on its upper wings. Their lifestyles, however, are virtually identical.

The mourner is America's Everydove, as familiar in eastern forests and suburbs as in midwestern grainfields or the grasslands, deserts, and mountain valleys in the West. Actually it's two birds—*Zenaida macroura caroliniensis* in the East and *Z. m. marginella* from the Midwest to the Pacific Coast—but only a skilled taxonomist can tell them apart. The birds themselves don't seem to care, especially in spring and summer when they're intent on turning out their annual contributions to the population.

The classic voice of the turtle is the male mourner's mating call—the perch coo, as it's known to ornithologists. It serves the dual functions of attracting a female and warning other males away from his territory.

What happens afterward depends on who shows up. If it's another male there's almost sure to be a fight, and no mere posturing or shoving match, either. During the breeding season male mourners are as belligerent as bull elk, especially if a female is present. They attack one another with outstretched feet and flailing wings. Being flogged by a dove, to say nothing of kicked, doesn't sound like much, but it's a slugfest if you're their size. One naturalist describes these conflicts as "bloody," which I doubt is very often the case, but they surely are feathery, partly because a mourner's plumage isn't very securely embedded in its skin and partly because these are serious fights that don't always end after the first round.

If a female shows up by herself it's another story. Then, the male goes through an elaborate courtship ritual involving bows and scrapes, tail-fanning, and graceful display flights. If she's impressed, they join in a round of cooing and preening and progress to billing and head-bobbing with her beak inserted into his. Copulation typically follows, and once the pair bond is formed, they'll stay together the whole season and possibly the rest of their lives.

Mourners are highly prolific and enjoy the longest breeding season of all North American birds, which is a cornerstone to their success as a species. A mated pair certainly wastes no time in starting a nest, nor do they waste any time in building a good one or being picky about where they put it. The male chooses the site, which may be on the ground, in a shrub or tree, a building ledge, even an abandoned nest built by some other bird. He also begins the construction, although the word hardly describes either the process or the flimsy, jerrybuilt jumble of twigs that passes for a dove nest.

Once the project is underway, he flies off to gather more sticks and, for reasons that defy explanation, typically lands on his mate's back when he returns. Perhaps it's the most convenient way of passing objects between them, as the Columbidae don't have very strong bills or strong feet for grasping. Or perhaps their combined weight helps compact everything. Or maybe they just do it because they can. At any rate he stands on her back, gives her the twig, and she sticks it into the growing pile. If he doesn't like where she puts it he'll give her a little peck under the chin; then she stands aside while he fiddles with it.

When they're satisfied that they have enough, the job is over. There's no lining, no particular shaping to form a cavity, just a platform of jackstraws often barely dense enough to keep the eggs from falling through. The whole mess is apt to be blown away by a strong wind, but if that happens it's no loss of fine craftsmanship and they'll slap together a new one just as quickly.

Two or three days later the nest will hold two eggs, which are incubated for two weeks. Both adults share this duty, the male during the day and the female at night. In some ways doves are more like songbirds than game birds. For one thing, hatchlings are altricial—which means they're blind, helpless, only sparsely covered with down, and monumentally ugly. All other game birds are precocial—that is, they hatch open-eyed, warmly dressed, and able to leave the nest within a few hours. They're also monumentally cute—bright-eyed little balls of fuzz with comically big feet.

Dove nestlings' first food is unlike any other birds' except other Columbidae—in fact more like that of mammals than birds at all. A few days before the eggs hatch the pituitary hormone prolactin begins to stimulate both parents, just as the same hormone stimulates milk production in female mammals.

As John Madson describes it in his splendid little book on the mourning dove: "The walls of the doves' crops thicken. Glands in the crop walls begin to secrete a whitish liquid that appears to be composed largely of fatty cells that are shed from the lining of the crop, just as such cells are shed from the mammary glands of a mammal. Pigeon milk is extremely rich in calcium and in vitamins A, B, and $B_1$. It's a creamy substance with a chemical makeup quite similar to rabbit's milk, and is highly nourishing. In less than two weeks, a squab will increase its weight almost thirtyfold. The male dove produces more pigeon milk than does his mate, and probably supplies more of it to the young doves than does their mother. In fact, male doves and pigeons are the only male creatures in the world that lactate for the purpose of feeding extremely young offspring."

Similar as the substances are, the feeding process is quite different between columbids and mammals. The adult opens its beak, the nestlings—often both of them at once—insert theirs, and the parent simply regurgitates the milk from its crop. After the first couple of days the mixture contains increasing components of insects, worms, and seeds, and after about nine days, less than a week before the young are fledged and out of the nest, the supply of milk ceases altogether.

But it will presently start again, because as soon as one pair of young are on their own the parents start over with two new eggs. Although overall nesting success is seldom better than about 50 percent—that is, only about half the total number of eggs result in fully fledged juvenile birds—a mated pair will stage four or five nestings in a season, perhaps even more in the most temperate climates. In southern latitudes, nesting mourning doves have been documented during every month of the year.

This naturally goes a fair way toward explaining why mourning doves are so numerous, and so does their feeding habits. Columbids are seed-eaters, and virtually any sort of small, firm seed—from cereal grains, weeds, grasses, flowering plants, whatever— is dove food. Doves do need water, which is scarce in arid regions, but the birds are fully capable of flying many miles for their twice-daily drinks. These they take by simply dipping their bills and sucking it up. So far as I know columbids are the only birds that don't have to tip their heads back to let water trickle down their throats.

Although they summer far into the north, mourners are warm-weather birds at heart. A few die-hards tough out the winter as far up as Puget Sound, the Great Lakes, and southern New England, but most of the northern birds start pushing south as early as August, spurred on by chilling nights and fall rains. The more robust white-wing is a pansy by comparison, eschewing cool

weather to such an extent that come January you won't find them even as far north as the Mexican states of Nuevo Leon and Tamaulipas.

Along with the mourners' southward trek comes dove season, which traditionally opens September 1. Because doves are migratory, season dates and bag limits are established by the federal government rather than the states—but the states can decide whether to host a season at all. Sixteen northern states from

Montana to Maine do not. Elsewhere, September 1 is a virtual holiday, especially in the South where dove shooting is a social as well as a sporting event.

And sporting it is. The mourning dove is the most popular game bird in the United States and certainly the most demanding of a gunner's skill, thanks to its speed and the breakneck stunts and chandelles it can pull with the merest prompting of alarm. Anyone who can bag a limit of mourners, taking the shots as they come, with a box of cartridges or fewer (that's fifteen birds with twenty-five shells) is an exceptional shot who's having an especially good day. I've seen many a good gunner run up strings of consecutive misses nearly to triple digits.

From what I've seen while hunting them in Mexico, whitewings aren't as fond of aerobatics as mourners are, but it doesn't seem to be any handicap to their survival. The first few I met, years ago in Sonora, didn't seem difficult at all, loafing in to feed in a cut grainfield, which led me to wonder about the stories I'd heard of their sportiness. But about then I noticed a pair headed my way, flying about four feet apart—a piece of cake as a right-and-left double. At thirty yards I started to lift my gun. They saw the movement and in the next instant were over my head and gone in a straight acceleration so blazing fast that they were out of range before I had time to turn around.

Wingwise, the South American eared dove falls somewhere between, but I cherish the hours I've spent with them, too. Doves delight me.

And they still puzzle me as well. The voice of the turtle is no longer a mystery, but watch a dove walking around in its waddly gait, constantly bobbing its head back and forth as it moves, and tell me: How must the world appear to a walking dove?

I've tried bobbing my head and walking the same way, but it just makes me go dizzy and bump into things. I'm resigned to the fact that I can never fly like a mourner, much as I'd love to, but it's bemusing that I can't even walk like one. There's more to this than meets the eye.

◆

# III

# Might and Mettle

# 12

# Lord of the Lakes

If I remember correctly, the first piece I ever wrote about any wild animal was a two- or three-paragraph affair on moose. As I was at the time an eighth-grade student in southeastern Iowa, it was not drawn from any great personal experience—though I had actually seen a moose in a zoo one time—but rather from a painting my English teacher had hanging on her classroom wall. What it had to do with a curriculum largely devoted to *Ivanhoe*, Heywood Braun, and *The Merchant of Venice* I haven't a clue, but there it was.

I recall the painting and the story well enough. The one showed a bull moose in profile, posing nobly upon a rocky promontory with a rugged mountain landscape in the background. The other was a youthful paean to wilderness, wildness, and nature's monarch—or some similarly drippy metaphor manufactured by a youthful imagination equally fascinated by nature and old-fashioned prose.

But my teacher liked it, bless her. She read it to the class, gazed out upon us and said, "Why, that's *poetry!*" I cannot for anything remember her name, but I will never forget the compliment. In some ways it has affected my life ever since.

The feeling I had for moose was genuine, even if its expression was a bit strained. For years I fancied I would begin my career as a hunter of big game by bagging a specimen of *Alces alces.* This notion's demise coincided precisely with my first actual view of a bull moose in the wild. Though still fairly young I had at least accumulated the good sense to realize that moose hunting undoubtedly is keen and demanding, but when three-quarter-ton of dead animal hits the ground in some alder swamp or tamarack bog a zillion miles from nowhere Sport instantly transubstantiates to the thing called Work. From there I resolved that my career as a hunter would not involve anything too large to comfortably carry home in one trip. But neither that nor the fact that I've subsequently had few opportunities to spend time in the company of moose has dimmed the fascination.

What's kept us apart is mainly geography. Moose don't live where I do. In fact moose don't live in much of the United States. In the East their range covers most of Maine and the far northern tips of New Hampshire, Vermont, and New York. Midwestern populations live on Michigan's Isle Royale and across northeastern Minnesota. In the West they live in the northern Rockies, ranging far enough south to edge into Colorado and Utah.

Otherwise, real moose country is farther north, in the belt of mixed hardwoods, boreal forest, and tundra that circles the top of the world. There they evolved and there for the most part have they stayed. In presettlement days, North American moose ranged a bit farther south but not much—into what now is Massachusetts and along the southern shores of Lakes Ontario, Huron, and Superior.

In Europe they're called elk, in North America moose, a name derived from one of the Algonquian languages. In taxonomic terms they are *Alces alces*. Current opinion recognizes six subspecies worldwide, three of which inhabit North America. The Canada moose is the most widespread and as the name suggests ranges all across Canada from the Maritimes to the Yukon and Alaska, where it's replaced by the Alaska or, to some, Alaska/Yukon moose. The third found in the western United States is called the Shiras moose.

Regardless of names, moose are deer, the largest members of the family Cervidae. They may in fact be the largest species of deer ever. I'm not sure how the modern article would stack up against *Cervalces,* the extinct, so-called giant moose, but it seems to be larger-bodied than the also-extinct Irish elk, whose antlers reached a spread of almost twelve feet. Of the New World animals, those of the Shiras strain are the smallest while the Alaska/Yukon moose are the largest—the largest in fact of any in the world.

All this of course is relative. There are no small moose. A mature Alaska/Yukon bull can measure ten feet from nose to tail—and at no more than two inches, a moose tail is a negligible quantity. He may stand seven feet or better at the top of his humped shoulders and weigh 1,800 pounds, which would make him a foot taller than a bull buffalo and half again heavier than the most robust brown bear. Cows are smaller, but even at an average three-quarters the size of bulls, they are by no means dainty.

In homelier terms, a bull moose is longer than some automobiles, weighs about as much as six refrigerators, and if he were standing in your living room, his shoulders would almost reach the ceiling—his *shoulders,* mind you, not the top of his head nor the tips of his antlers. A couple of years ago I was running up the Copper River in Alaska one morning on a fishing trip, when the guide zipped the jet boat around a bend and almost ran under a moose standing hock-deep in midstream. I really think we could've passed under his belly without touching, he looked that big. Fortunately he made a couple of startled leaps, cleared the seven-foot bank in one hop, and disappeared into the streamside shrubbery, leaving me with a vision I won't soon forget.

Everything about moose is oversized. Their hooves are wide and splayed to support a heavy body on snow and muskeg and marsh. Their ears, which they can cock in different directions at the same time, are huge sound-scoops adapted to forest habitat where keen hearing is more useful than sharp eyesight. Their muzzles are almost grotesquely long, allowing them to feed in water and still keep their eyes above the surface. And a bull's massive, palmated headgear is easily the most impressive adornment in antlerdom.

An average bull in his prime wears antlers with a four- to five-foot spread; on a really big one the span can be six feet or more. At eighty to a hundred pounds per pair, they are the heaviest headwear—antler or horn—grown by any animal on earth.

And the marvel is that every bull grows a new set every year. As a yearling they are simple spikes. His second set are odd-looking little devices commonly known as "shoehorns" or "bootjacks," and only in the third year does the typical mooselike shape emerge. From then, each set will be larger until he reaches the age of eight or ten, at which time his antlers will be as big as they're ever going to get.

Regardless of size the whole process takes only about fourteen weeks, beginning in April when the buds start to sprout from the pedicels on his skull. The growth ends in July. Like those of all cervid animals, moose antlers grow in velvet, which is a thin layer of skin rich in blood vessels and furred with soft, fine hair. When growth reaches the maximum, blood flow in the skin subsides and the material underneath begins hardening to the consistency of bone. Presently the skin dries and cracks and for a while hangs in rags and tatters as the bull thrashes trees and brush, practicing behavior he hopes will allow him an active role in breeding.

Whether it does depends upon his age and size and probably some other factors known only to moose. If he's less than three or four years old his chances are slim; where populations are especially dense he may be as old as five before he can hold his own against older bulls and at the same time sufficiently impress a receptive cow.

One thing for sure: As a love-object a cow moose could only be attractive to a bull moose. His antlers give him a considerable measure of handsome dignity, even grace. As cows don't grow antlers, she has nothing to alter the fact that she looks like the product of a union between a mule and a stunted giraffe. Gawky, angular, and ungainly, an antlerless moose isn't likely to win any beauty contests—an appearance unaided by the loose pouch of long-haired skin both sexes wear at their throats. Commonly called the "bell," this device serves no practical purpose that anyone's been able to figure out.

Odd as they may look to everyone else, though, bulls and cows take a definite shine to one another every fall. The season begins in mid-September and lasts through October. Bulls normally are peripatetic during the rut, seeking out cows wherever they happen to be and dealing with other, competing bulls as they meet. But in areas where the population is unusually dense—say, as many as two moose per square kilometer—dominant bulls often establish mating territories.

Either way they settle questions of pecking order in much the same manner—intimidation displays of brush-flogging, posturing, and a vocabulary of grunts and groans. Sometimes the confrontations evolve into shoving matches and even full-scale combat, but as all-out fighting does the overall mating objective more harm than good, things rarely go that far.

The classic notion of moose during the rut has it that cows announce their availability with ringing bellows, sounds that a lot of moose-hunting guides imitate to attract bulls. After a great deal of observation, however, Canadian artist and moose-fancier Gisele Benoit suggests the reality to be just the opposite—that cows bellow when they are sexually unreceptive, the alced equivalent of "Don't bug me, Charlie; I'm not in the mood." Since a cow's actual period of fertility lasts only about twenty-four hours, it makes sense that she might be interested in hanky-panky only when it's likely to count.

Bulls being bulls, on the other hand, the old guide's trick often works, simply because a bull moose in rut responds to any cow he can find, regardless of whether she may ultimately put him off. Considering that he prepares for amatory adventures by making a wallow of his own urine, his general lack of seductive delicacy isn't surprising. Besides, her response won't be a total shut-down because she'll want him there when the time is right. He therefore will hang around for perhaps a week, do his part in aid of the next generation, and then go looking for another cow.

Predictably for an animal so large, gestation is a lengthy process—240 to 250 days—and the calves are born in late May and early June. A cow's first pregnancy, which occurs when she's two or three, produces only a single calf, but in subsequent years she frequently will bear twins and on rare occasions even triplets. The calves weigh about twenty-five pounds each and are able to walk and suckle within a few hours of birth.

But between breeding and birthing seasons lies winter, and if northern winters aren't quite as hard on moose as on the smaller deer, they still emphasize the problems of survival.

Like all deer, moose are vegetarians and cud-chewers, equipped with the standard multichambered digestive system designed to make the most of marginal fare. As they are browsers rather than grazers—feeding on twigs, leaves, buds, bark, and a variety of plants instead of grass—they must forage more or less continually to maintain adequate levels of nutrition. Willows, maples, mountain ash, balsam, birch, and aspen are mainstays, but whatever a moose eats, it needs a lot of it. Stoking a body so large requires fifty to seventy-five pounds of food every day. By comparison, a half-dozen white-tailed deer can live comfortably on the same amount.

But moose are well equipped to meet such demands, so long as their numbers remain in balance with the available food supply. Their height gives them a tall reach; long legs and big feet allow them to move about in snow that would stall a deer or even an elk. A foot and a half of snow is enough to send whitetails yarding up on some sheltered spot; it takes thirty inches or more before moose will do the same.

In summer moose relish nothing so much as a meal of aquatic plants, which they can harvest at any depth from a few inches to almost twenty feet. They are the only members of the deer family able to submerge completely and can stay underwater for nearly a minute at a time. All of the cervids can swim if they need to, but only moose seem willing to take to water for some reason besides getting to the other side. It's no accident that one of the classic scenes in wildlife art combines moose with the wild beauty of a north-country lake. As you might expect, they're extremely powerful swimmers able to outdistance all but the hardiest canoe paddlers, and they can keep up the pace for ten miles or more.

Moose can be remarkably difficult to approach, especially where the population has been chivied by man. In the old days the Crees and Chipewyans could pay no higher compliment than to say that a man was a moose hunter. As Ernest Thompson Seton put it at the turn of the century, "Moose-hunting by fair stalking is the pinnacle of woodcraft...At Fort Smith [on the Slave River in the Northwest Territories] are two or three scores of hunters, and yet I am told there are *only three moose-hunters*." The italics are Seton's, but the admiration belonged to almost everyone.

I'm willing to take Seton's word on how demanding moose might be as game, having chosen long ago to be a hunter exclusively of birds. As another factor in that decision, apart from the drudgery endemic to big game, I have always resolved to eat at least a fair portion of everything I shoot. With small game I may not care for the taste, but at least I don't have half a ton of it on my hands. I might be persuaded to make an exception for *Alces alces*, though. I've eaten a lot of moose meat, and it's peerless fare, every bit the equal of any haunch, roast, or steak you can carve from an elk, a whitetail, or a buffalo. Even moose-nose is great stuff when it's prepared by someone who understands the vagaries of cooking game.

Man isn't the only creature with a taste for moose, although only a few predators are capable of doing anything about it. Bears and mountain lions kill a certain number of calves and perhaps even some yearlings. The most important natural predator is the wolf, who by dint of numbers and magnificent hunting skills can take full-grown moose as well as calves. This is not, however, to say that wolves kill every moose they find, nor that moose populations are in any danger from wolf predation. Studies conducted in Isle Royale National Park prove beyond question that wolves and moose can coexist perfectly well.

Actually parasites and diseases kill far more moose than man and predators combined. Like all ungulates, they are vulnerable to actinomycosis, necrotic stomatitis, a variety of bone and gum diseases, and hoof infections. Winter ticks often parasitize moose in incredible numbers, virtually bleeding them dry. Warble and nostril flies attack them as well. Golfball-sized hydatid cysts, which house the larvae of certain tapeworms, frequently form in their lungs; biologist David Mech reports finding fifty-seven of these in the lungs of one wolf-killed moose. Under the circumstances, one might argue that the wolves performed an unwitting act of mercy.

Though death in nature never is a pretty sight, no beauty can compare with that of natural life, and something about a bull moose in his prime far transcends his ungainly aspect. As H. E. Anthony puts it in his 1928 *Field Book of North American Mammals*, "Although it is almost a caricature of the graceful forms of the smaller Deer...there is a suggestion of massive strength and irresistible vigor about a Moose that is certain to arouse a feeling of admiration."

True enough—which may be why *Alces alces* is the only animal ever to be the namesake of an American political party. This is not to overlook the so-called Bucktails, who were active in New York State politics from 1816 to 1830, but they were a splinter faction, not an official party. The organization I have in mind is the Progressive Party, under whose aegis Theodore Roosevelt ran unsuccessfully for the presidency in 1912. During the campaign a reporter asked Teddy how he felt, and the old Rough Rider replied, "I feel as strong as a bull moose." From then until the Progressives disbanded in 1916 they were known as the Bull Moose Party.

Physical power is one thing, imaginative association quite another. Like the sound of a loon, the sight of a bull moose insouciantly browsing the shallows of some remote northern lake prompts all sorts of metaphors that we'd otherwise do well to avoid. I fancy I have a better view of wild things than the one I put into my little essay thirty-five years ago, but still I find it hard to entertain an image of majesty that doesn't have a moose mixed up in it somewhere.

# 13

# The Beast That Grazed the Lawns of God

On Thanksgiving Day, 1925, Charles M. Russell sketched a lone American bison at the top of a sheet of foolscap and filled the rest of the page with a letter to Ralph Budd, president of the Great Northern Railroad.

"turkey is the emblem of this day and it should be in the east," Russell wrote, spelling by ear and exercising his lifelong indifference toward punctuation, "but the west owes nothing to that bird but it owes much to the hump backed beef in the sketch above

"the Rocky mountains would have been hard to reach with out him

"he fed the explorer the great fur trade wagon tranes felt safe when they reached his range

"he fed the men that layed the first ties across this great west Thair is no day set aside where he is an emblem "the nickle wears his picture dam small money for so much meat he was one of natures bigest gift and this country owes him thanks"

Russell was right, as far as he went. Although the beaver played a more significant role in the exploration and settlement of North America, no animal stands closer to the heart of history and prehistory on this continent than the bison. Almost from the moment the first pioneers left the Atlantic at their backs, the United States owed the bison thanks—and by Russell's time, owed him a profound apology as well.

In presettlement days, bison ranged virtually throughout North America, from Alaska southeast to the Great Lakes, east nearly to the coast, south around the Gulf to beyond the tropic of Cancer, west to the coastal mountains. Overlay this former range on a current map and no more than six or seven states, in New England and on the upper East Coast, lie outside it. Only the wolf was more widespread.

The same map will show cities, towns, or villages named Buffalo in fifteen states. Four—Minnesota, Iowa, South Dakota, and Texas—have more than one community of which *Buffalo* is part of the name. Get closer to the ground, and you'll find dozens of rivers, lakes, streams, forks, canyons, mountains, springs, passes, meadows, hills, valleys, ridges, and roads all named for the great shaggy beasts. In its impact upon North American place names, no other animal even comes close. Many a modern highway, particularly in the Appalachians but elsewhere as well, follows the routes of ancient buffalo roads.

No matter how good your map, however, you'll be hard-pressed to find much of anything, whether community or local geography, named Bison, although technically that's the proper name—*Bison bison*, kin to all the bovine animals of the world, from goats and sheep to cattle and antelope and wildebeest, the true buffaloes, the bushbucks, nilgai, oryx, and the rest. Call them bison if you wish, but to the American consciousness, they are buffalo. Besides, *bison* is too clipped and prissy sounding to capture any sense of the beast itself; that's hard enough even for the fine, rolling syllables of *buffalo*, but it's a better start.

Curiously, more than half the states with towns named after *Bison bison* are east of the Great Plains, where populations never were particularly large or dense. Even the states where the magnificent tallgrass prairie flourished—Minnesota, Iowa, Illinois, Missouri—held relatively few buffalo compared with the mid- and shortgrass plains that reached from Canada to Texas and west to the Rockies. There, explorers and pioneers found them in an abundance almost unimaginable.

The buffalo did not originate in North America. Its ancestral forms came from Asia, migrating across the Bering land bridge in at least two great waves. The first began about 50,000 years ago and lasted about 10,000 years, until glaciers blocked the way. Climatic shifts reopened the route about 12,000 years later, and it remained so until about 10,000 years ago. During both periods, Asian animals made their way east into the New World while such North American natives as horses and camels drifted the other direction.

Some of the ancient buffaloes naturally remained in the Old World, where they evolved into the present-day European wisent, a somewhat smaller, less hump-shouldered cousin of the American bison. Those that crossed the Bering bridge went through at least eight definable forms before becoming the buffalo as we know it. The largest, *Bison latifrons*, was an enormous beast, its horns nine feet from tip to tip. It was succeeded by the smaller *B. antiquus figginsi*, which was hunted in North America by the ancient men who chipped the graceful, fluted stone blades we know as Folsom points, and was itself replaced by the smaller-horned *B. bison occidentalis. Occidentalis*, in turn, evolved into the modern form about eight thousand years ago.

Buffalo adapt about equally well to a livelihood of grazing or browsing. Indeed, like the wisent, some of the American animals clung so tenaciously to an ancestral preference for living in woodland and feeding on leaves and twigs that they eventually were recognized as a distinct subspecies—*B. b. athabascae*, the wood bison, somewhat larger, darker, and woollier-coated than the plains buffalo, *B. b. bison*.

*Athabascae* may be bigger, but the plains buffalo achieved the triumph of sheer numbers. In the roughly 1.25 million square miles of grassland between the eastern forest and the Rockies, the land one writer described as "the lawns of God," the buffalo prospered almost beyond belief. No one will ever know how many there were when European man first arrived. Early observers compared their numbers to "the locusts of Egypt" and the fish in the sea, spoke of continuous herds reaching from horizon to horizon—or as one man put it, "The country was one robe." Estimates run as high as 60 million or more. Half as many is far more realistic, but even at that, 30 million large animals of a single species is a spate unlike any the modern world has ever known.

In pristine condition, such a vast environment populated with so many creatures would be a paradise for men whose numbers were few, and so it was for the Plains Indians, nomadic and semi-nomadic tribes whose territory lay in a great swath through the center of the continent, from Canada to the southern plains—the Sarsi, the Blackfoot, Assiniboin, Gros Ventre, Crow, Hidatsa, Mandan, Arikara, Yankton Sioux and Teton Sioux and Santee Sioux, Arapaho, Ponca, Pawnee, Oto, Cheyenne, Iowa, Kansa, Missouri, Osage, Omaha, Kiowa, Kiowa Apache, Comanche, and Wichita.

In the Plains Culture, the buffalo was everything from provender to god. He was *pte*, so infinitely useful, so profoundly the source of physical and spiritual livelihood that they called him Uncle. Every scrap of his body served some purpose. They ate his flesh and his marrow, made clothing and tipis and saddles from his hide, sled-runners from his ribs, ladles and cups from his horns, rope from his hair, blades and brushes and needles from his bones, thread from his sinews, vessels and pouches from his intestines and scrotum, holy charms from his great skull. On the plains, nothing of *pte* went to waste except perhaps his angry bellow. The buffalo was a character of myth, an image of spirit, object of worship. He was the incarnation of life itself.

Never has a human culture intertwined so deeply with a single animal. Even now, he remains the quintessential symbol of the American West.

Big land, big animal. A fully grown plains bison bull may stand six feet or better at the shoulder, measure ten feet from nose to rump, and weigh from 1,400 to 2,200 pounds. His horns may be a yard from tip to tip. Wood bison are larger still, taller both at the hump and in the hindquarters; a big bull might weigh 2,500 pounds. The cows of both subspecies are smaller, bearing only about half the bulls' weight on a frame that's a foot or more less in all dimensions. They're big enough, even so. Having spent some time with both, I can tell you that standing among a band of grazing buffalo makes dairy cattle seem like a herd of puppies by comparison.

Zoologists have largely discredited the old notion of a great annual buffalo migration, but still the herds were more or less always on the move, drifting hither and yon across the plains, influenced by the cycles of weather and grass.

◆

In a way, the grasses created the buffalo, especially the hardy plants that thrive in the broad rain-shadow of the Rockies. It is a land of extremes, where fierce cold alternates annually with blazing sun and where rainfall is erratic. There, grasses thrust farther below the surface than above it, surviving on a root system that might reach nine feet or deeper. The blades of all grasses grow from the base, not from the tips, which makes them highly resistant both to fire and grazing. Indeed, cropping and burning stimulates some species into even more luxuriant growth. Although grass is not particularly nutritious, it grows in abundance, and to exploit such riches, buffalo and other grazers evolved a four-chambered digestive system efficient enough to support the animal while still leaving enough nutrients in the byproducts to act as fertilizer. Grazers and grassland thus form a mutually sustaining cycle.

The buffalo's senses also evolved in accordance with its life as a vegetarian and a dweller of open spaces. Although its vision is sharp enough to distinguish a mounted horse from one without a rider at half a mile or more and to perceive moving objects at distances greater still, hearing and smell are the animal's main tools both for defense and for keeping track of one another.

Although bison herds often were vast in the aggregate, the infrastructure comprises smaller groups—family units under the leadership of a cow, larger societies of cows and young, and bachelor groups of bulls. For most of the year, they maintain these groups within the larger context of the herd, content to pursue peaceful lives, grazing and sleeping and indulging in bisonine pleasures.

The keenest of these seem to center on touch. Buffalo love to rub their heads and necks and flanks—against trees, bushes, stumps, rocks, stream banks, even the mounds of earth at the entrances of prairie dog burrows. In short, a buffalo is likely to rub against anything that protrudes, which in the early days brought a measure of grief to some who built log cabins and soddies where the buffalo roamed. It also did nothing to endear the animals to the crews who strung the first telegraph wires across the plains.

Apparently, rubbing just feels good. It does serve some grooming function, especially in the spring when buffalo molt their heavy winter hair, but wallowing does even more to keep them tidy. Lying on some patch of bare ground, they thrash and wriggle, raising clouds of fine dust in the not-inconsiderable effort to roll onto their humps before their huge bodies come thumping back down. The friction peels away molting hair in rags and tatters, while the dust acts as both cleanser and treatment against parasites. Generation after generation of animals used the same wallows, wearing out bowl-like depressions you can still see all across the plains, wherever the land hasn't been plowed. The patient, healing grass has grown over them in the nearly 120 years the buffalo have been gone, but you can find them if you know what to look for.

In midsummer, from the end of June to early September, the buffalo's social structure changes. With the onset of rutting season, the herds grow restless. Small knots of cows drift together and form larger groups, toward which the bulls gravitate as their own bachelor herds break up.

I've never been very close to a bull buffalo in rut, and to be honest with you, I don't care to be. His bellowing, ground-pawing, head-butting, horning, wallowing (which only bulls do at this season), and generally roughneck behavior is impressive enough from a respectful distance. Even in the comparatively minuscule herds that exist today, a few bulls can set up a frightful clamor; it's hard to imagine the magnitude of chaos they must have caused when the populations were at their peak.

Bulls establish mating rights mostly by intimidation and bluff, occasionally in a pushing match, and once in a while in a vicious battle that looks and sounds like the clash of titans. For all their ponderous bulk, they're astonishingly agile and quick. They're fast on the hoof—thirty-five miles per hour or better—possess remarkable endurance, and can turn in what appears to be little more than their own length. A dust-raising, sod-ripping dispute between two bulls primed for the rut is no minuet.

Once they have things sorted out among themselves, they still have to deal with the ladies, and that isn't always easy. If a cow is not at the proper point of her estrus—or otherwise not in the mood—she'll rebuff any amorous advance, sometimes by a gouge from her horns and sometimes with a kick that could fell anything from a grizzly bear to a Volvo. The bull on the receiving end, his brain and body awash with hormones, scarcely seems to notice. Eventually, of course, the cows of breeding age relent. Those not bred the first time around come into estrus again about three weeks later.

Bulls maintain their mating frenzy throughout the rut, and this can prompt them to lose their heads altogether. About thirty years ago, I had a good friend who spent his summers working as a ranger in Rocky Mountain National Park, which had a free-ranging buffalo herd, and he came home one fall with a story that still makes me smile whenever I think of it.

Despite the signs posted to warn visitors against approaching the buffalo or driving anywhere off the roads, some dimbulb in a Volkswagen Beetle decided to have a closer look and went trundling off across the prairie toward a distant band. It was late summer, peak of the rut, and the resident boss bull perked right up. He probably had bred every available cow in his little herd by then, so maybe he was feeling a bit frustrated. At any rate, he took a definite shine to the Bug and in the grips of passion managed to cave in the top, break the driver's arm, smash the rear window, and in general reduce the thing almost to rubble.

How'd you like to explain *that* to your insurance company?

In aboriginal days, such distractions naturally did not exist, and with a superabundance of cows to attend, the bulls probably were ready for a rest when the mating urge began to subside in mid-September. The cows, newly bred and otherwise, went back to their matriarchal clans, the bulls to their bachelor groups, and relative calm returned to the prairie.

Gestation takes about nine months, through the winter and into the softer days of spring. In April and May, oblivious to the gorgeous spectacle of the prairie in bloom, the heavily pregnant cows begin wandering away from the herd to perform the solitary act of birth. It will result in a single calf; buffalo twins are extremely rare.

Delivery may take a few minutes or as long as two hours, but when it's complete, the cow immediately devours the fetal membrane and then spends half an hour licking her calf, doing both in an instinctive effort to replenish her body's nutrients. For the same reason, she also will consume the placental mass when it's expelled a while later.

A newborn buffalo calf is a gawky package, all gangly legs that at first refuse to work. Within a couple of hours it will learn to stand and take enough spraddle-legged steps to reach its mother's udder and begin nursing. As for a great many animals, the first few hours of a buffalo's life involve a process of imprinting in which the calf comes to associate with something in its environment. As a survival mechanism, this ensures a sufficient level of parental care, since the young animal will follow whatever it associates itself with. The proper object, of course, is its mother, which no doubt is why the cows go off to give birth alone. But buffalo calves have been known to imprint upon other members of the herd, horses, and even people. Those that do not form the connection with their mothers will become orphans, straggling aimlessly behind the herd as it moves. Occasionally they are adopted by another cow; more often they starve or become a meal for some predator.

Not surprisingly, only very young buffalo are at any great risk from predators—and from a fairly short list of them, at that. In aboriginal times, more buffalo probably fell to wolves than anything else, followed by coyotes and grizzly bears. Golden eagles probably accounted for a few, and mountain lions no doubt took some from herds in mountain parks and valleys. Nonetheless, given the buffalo's numbers and gregarious habits, predation never was significant even where predators were relatively plentiful.

By three months, about the time the rutting season begins, the young of the year are both secure and well integrated into the life of the herd. Their natal coats of red-orange have molted, replaced by the dark-brown pelage of the adults. They still supplement their diet of grass by suckling, but this grows increasingly infrequent, until they are fully weaned at about nine months. In the first year, the calves increase their thirty- to forty-pound birth weight by a factor of ten. When their mothers give birth again, the yearlings are on their own for good.

They are, however, members of the herd, living cogs in a homogeneous mass that can sometimes act with the purposefulness of a single organism. Leadership among any herd animals exists only in relation to the degree of followership demonstrated by the throng. This is evident on even a small scale; in the old days of huge populations, it must at times have been an awesome phenomenon. Pioneer journals describe buffalo stampedes of almost unimaginable proportion. What began as a flight from danger, real or only perceived, quickly took on a life of its own, spreading through the herd like a chemical reaction feeding upon itself, until the whole landscape thundered with wave after wave of animals rushing nowhere in a blind frenzy. In the crush, they simply overran whatever lay in their path. Whole wagon trains were trampled flat, steam locomotives toppled from their tracks, steamboats battered and swamped as the herds swam the rivers.

Charlie Russell's estimation of the buffalo's contributions to the westering pioneers is certainly true, but they did not come without a price. The animals fouled waterholes beyond human use, which often was a great hardship in country where waterholes were few to begin with. More than a few parched travelers owed their lives to the liquids drained from the paunch of a freshly killed buffalo.

Those who simply passed through buffalo country found the animals to be a vital, if not altogether benevolent resource, but those who sought to remain on the plains found them difficult neighbors. Homesteaders and ranchers were powerless to protect their struggling crops and grazing kine against the herds that ate and trampled the one and dispersed the other beyond recapture. Finally, the whole matter came to a head.

In the years following the War Between the States, America cast an increasingly greedy eye toward the frontier. Expansionist dreams, myths of limitless land, utopian ideals, and a few other basically conflicting instincts all pointed west. Fueled by bandbox rhetoric and classical philosophy, envisioning a vast nation stretching from one ocean to another, the newly reunited union determined to claim whatever might lay within its grasp. That the Great Plains—in fact, the entire West—was already occupied scarcely mattered; America needed the land for its own people and the grass for cattle. Under the European notion that everything must belong to someone, Manifest Destiny could hardly be denied.

If the buffalo were a problem, the Indians were an even greater one—although it was plain to see that an animal and a race of people so intertwined in life must naturally be susceptible to a common fate.

White men had been hunting buffalo for a generation or more, both for sport and as provision for the workers building railroads and the telegraph. Indeed, William Cody earned his nickname Buffalo Bill during the eight months of 1867-68 when he worked as a subcontractor supplying fresh meat to the Union Pacific railroad crews in Kansas.

Although the kills often were excessive, the buffalo were simply too numerous to be much affected by either sportsmen or meat hunters—nor, for that matter, by the trade in buffalo robes that thrived throughout the nineteenth century. By 1871, however, the tanning industry had developed new methods of treating hides, and markets in both the United States and Europe were hungry for leather.

◆

At the same time, the American government perceived a simple solution to the main obstacle of Manifest Destiny. As I said, recognizing the Plains Indians' profound dependence upon the buffalo took no great intelligence, and in the early 1870s, the power to do something about it rested with two men who possessed exactly that—President Ulysses Grant and his Secretary of the Interior, the fatuous and corrupt Columbus Delano. The Bureau of Indian Affairs was organized under the Interior Department, and in his annual report of 1873, Delano wrote, "I would not seriously regret the total disappearance of the buffalo from our western prairies, in its effect upon the Indians, regarding it rather as a means of hastening their sense of dependence upon the products of the soil and their own labors."

Grant evidently agreed, since a year later he disposed by pocket veto of the first legislation attempting to bring the harvest of buffalo under some control. Subsequent bills simply died in Congress.

Thus, on tacit approval from Washington, the hidemen doubled and redoubled their efforts, and knowingly or not, they trained their rifles straight at the heart of the Plains Culture. The great herds soon were gone from Kansas to the Dakotas. The venue then shifted to Texas, to Montana, even into Canada, wherever remnant populations could be found. Everywhere across the plains, the countryside reeked with the stench of rotting carcasses, and even now, as a chapter in the history of American wildlife, the 1870s still bear the stench of shame. By 1883 it was all but over, and by 1889 William Hornaday of the National Museum could document no more than eighty-five free-ranging buffalo in the entire United States. Fewer than five hundred more lived in various parks and zoos and private herds, and about five hundred were known to exist near Great Slave Lake in Canada. The rest were no more than calcified bones scattered over half a continent. The Indians' day was all but finished as well.

Obviously, the buffalo did not disappear altogether, thanks to the efforts of a few early conservationists. Management programs that now exist ensure their continued survival, although certainly not in anything even remotely like their former numbers. What happened was inevitable, because 30 million animals the size of buffalo could not conceivably exist in a heavily populated, industrialized, agriculturalized country. Reprehensible as the slaughter may have been, we are hardly in a position to judge too harshly, living as we do upon all the fruits of the civilization in whose behalf the buffalo died. We couldn't have wheatfields and oilfields and communities connected by superhighways—all that and 30 million buffalo, too. And we're no more willing to give up these things now than our great-grandparents were willing to give up the promise of them a hundred years ago.

So long as it shares the planet with us, no form of terrestrial wildlife larger than insects is likely to ever have more than a modicum of space, and that reality, in turn, lends a certain clarity to the future, demands of us both the moral fortitude to grant the space and the intelligence to recognize that species are more important than individuals. To live without 30 million buffalo is one thing; to live without any at all is something else again. Perhaps the best thing we can learn from the buffalo is to know the difference.

◆

# 14

# Ancient Myth, Modern Magic

In the beginning, an ancient Zuni legend tells, monsters ruled the earth. One of them, a great antlered beast whose neck was a half-mile long, stood in the east and ate the clouds as they formed. Great drought afflicted the people.

Two brothers, the *Ahaiyute,* went off from their home on Corn Mountain to seek Cloud Swallower and with the aid of Gopher killed the beast with arrows. They cut open his breast and flung his heart toward the east, where it became the morning star. Cloud Swallower's liver in turn became the evening star, his lungs the Seven Stars, and his entrails the Milky Way. "Now," the *Ahaiyute* said, "there will be rain."

It isn't clear whether the original mythmakers had deer or elk in mind, but many North American tribes sought to promote fertility and rainfall through rituals in which dancers dressed in deerskins and wore antlers upon their heads.

How the Indians came to associate deer and rain no one knows. Both were elemental components of the world they knew, and that may be explanation enough.

Deer always have played a fundamental part in the life of man. They figure prominently in the mythologies of the Hittites, the Greeks, the Celts, the Norsemen, and others. Early Christians sometimes used the image of a deer to symbolize Christ. In Scottish folklore deer are often called "the fairies' cattle." They are not always benevolent characters in Indian myth, as the Zuni story indicates, but they almost never are cast in the role of the Trickster, either. Indeed, throughout the Northern and much of the Southern Hemisphere as well, deer are universally viewed as contributors to the good of mankind.

Exactly which animals locally enjoy such regard depends upon where you look, for the deer are a widely varied family that in one form or another inhabits every continent except Antarctica. As a group they belong to the order Artiodactyla, animals that walk on two toes in the form of cloven hooves. As a family they are the Cervidae, represented worldwide by thirty-eight different species in seventeen genera.

Although no modern deer is quite so spectacular as *Megaloceros*, the now-extinct Irish elk whose great, palmated antlers measured eleven feet or more from tip to tip, the present Cervidae come in a remarkable range of sizes—from moose to the raccoon-sized pudu of South America. The pudu stands about a foot tall and seldom weighs more than fifteen pounds. Between lie the musk, water, tufted, axis, red, and barking deer of Asia; the fallow deer of Europe, Asia Minor, and north Africa; the roe deer of Eurasia; the marsh and pampas deer and the guemals of South America; the brocket deer of Central and South America; the caribou of the circumpolar tundra; the North American elk, moose, mule deer, and the familiar whitetail.

The geologic record suggests that the family originated in Asia some 35 million years ago, soon migrated westward to Europe, moved eastward across the Bering land bridge into North

America about 25 million years ago, and from there eventually colonized South America. In historic times man's proclivity for transplanting his favorite animals hither and yon has greatly expanded the deer's horizons. As a result of deliberate introductions dating at least as far back as the Phoenicians and Romans, wild deer populations of one species or another now occur in Cuba, New Guinea, Australia, New Zealand, various islands in the Pacific, and in other places far beyond the family's original, natural range.

Cervids are able to thrive in nearly every habitat from forest to tundra, mountains to desert, grassland to brush country. But wherever they live and whatever their size, all deer share certain physical characteristics. They are ruminants, for one thing, cud-chewers of great appetite, and are therefore supplied with multichambered digestive systems. With two exceptions, all male deer grow antlers, and so do female caribou. Instead of antlers both sexes of musk deer and Chinese water deer grow tusks, saber-shaped upper canine teeth that may be nearly three inches long in the males and slightly shorter among females. The muntjac or barking deer of Asia and the South American guemal have both antlers and tusks.

Nearly all cervids are equipped with various scent glands located on their faces or their legs, or on both. Relatively few species form large herds, but most associate in loose groups and scent seems to be an important means of keeping in touch with others of their kind.

Though they have long lived in close proximity to man there are no domestic strains of deer—although the Lapps of northern Scandinavia have succeeded in partially domesticating some of the indigenous reindeer, as caribou are commonly called in Europe, to the extent that the animals allow themselves to be herded and sometimes are even willing to serve as beasts of burden. There is some evidence that the ancient Lapps achieved a similar relationship with moose (which in Europe are called elk).

The Lapps, however, are the exception, for man elsewhere has always regarded the various deer as wild game. Of all the animals sacred to Diana, Greek goddess of the hunt, deer held an especially valued place. Hunting deer with horse, hound, and longbow was a formalized sport in Europe by the Middle Ages. To the native inhabitants of North America deer were a staple source of food and various materials of daily life.

By the time early man reached North America, along the same route that brought deer into the New World, the ancestral Asian animals had evolved into two distinct forms—the big-eared, big-antlered mule deer of the West, and the slightly daintier, more widely spread whitetail, whose range extends virtually throughout North America, south to Peru and northeastern Brazil.

Despite some physical differences, especially in the way their antlers branch, the two are roughly the same size and lead similar lives. Both are classed in the genus *Odocoileus*, a name made of two Greek words that combine to mean "hollow tooth," referring to characteristic depressions in the molars. Mammalogists know the mule deer as O. *hemionus* and the whitetail as O. *virginianus*. All other deer of the United States and Canada—the blacktail, Columbian whitetail, Coues, key, Sitka, and Cedros Island deer—are subspecies of one or the other.

Deer are big game but not as big as many people, deer hunters included, often think. Size and weight vary widely, depending to some extent upon where the animal lives. Northern deer tend to be larger than southerners, and the subspecies tend to be the smallest of all. The elfin key deer, a whitetail subspecies found on only a few of the Florida Keys, stands about two feet tall at the shoulder and seldom weighs more than thirty pounds.

The typical North American whitetail will stand perhaps forty inches at the shoulder, measure about five feet from nose to tail, and weigh in at about two hundred pounds. Mule deer tend to be a bit larger.

Deer are dwellers of edge habitat and prefer their world to be a mixture of dense vegetation for concealment and open space where they can browse on the leaves, twigs, and fruits of an astonishing variety of trees, shrubs, grasses, farm crops, fungi, mosses, and lichens. White-tailed deer prefer acorns above all other foods and are extremely fond of poison ivy. If necessary they'll eat snails and even fish in small amounts.

Twilight and early morning are the prime feeding periods. When the moon is full and the sky is clear they may feed through the night, but the sun won't be far above the horizon next morning before they head back into their favorite thickets—a farmland woodlot, a tract of suburban forest, a patch of brush or tall grass. There they spend the day working over the food they've gathered.

Because much of their food contains relatively little nutriment, deer and other plant-eating animals need to put away large quantities of it, perhaps as much as a ton in a year's time. Owing to their natural status as prey for the meat-eaters, the more time they spend feeding in the open, the higher the odds that they will become someone else's dinner instead. To help solve both problems a deer's digestive system has four separate chambers—three specialized branches of the esophagus and one true stomach.

Feeding is little more than packing in a supply of raw material. Since deer have no upper incisors they cannot nip and nibble but rather pull plant matter into their mouths with a long, agile tongue, shear it free with their lower incisors, give it a perfunctory chew, and swallow it whole. It goes first to the rumen, the largest of the esophageal chambers.

The rumen is mainly a storage chamber. It does little or no processing, although it does contribute a vital quantity of bacteria before sending the bulk food into the second chamber, the reticulum. There, with the help of the bacteria whose job it is to break down crude fiber and cellulose, the material is softened and condensed into small plugs or cuds.

When the animal has gathered its fill it lies up in some protected spot for a few hours. During this time it draws the cuds back up its throat one by one, and at that point, well after feeding is over, the actual eating begins.

Deer wear an impressive set of molars in both upper and lower jaws, ridged and pleated grinding tools that finely shred even the most fibrous twigs and leaves. Once the cuds are thoroughly chewed and mixed with saliva they pass back down the throat and through a special valve in the side of the esophagus into the third chamber, the omasum. The omasum in turn sends the cuds into the abomasum, the true glandular stomach, which contributes gastric juices and passes everything along to the intestine.

Once in a while some indigestible object—a hard-shelled nut, a stone, a nail, or something else—gets into a deer's rumen and stays there. The rumen, perhaps in response to the irritation, secretes calcium that coats the object layer upon layer, just the way a mollusk builds a pearl. In a deer these stomach stones may grow to be two inches or more across, but even the smallest are traditionally precious items.

Generally they're called bezoars, a name derived from a Persian word denoting the expulsion of poison, and in many parts of the world bezoars have since ancient times been prized as sovereign remedies against poisoning. In American folklore they're called madstones and are believed to cure anyone unlucky enough to be bitten by a rabid dog. According to tradition a madstone will stick to the wound if rabies is present and will neutralize the disease if left attached for a certain time.

Madstones have always been valuable (one was sold in Texas in 1879 for $250) and some have become famous, their pedigree of cures carefully noted as they've been passed along through the generations. The Nelson madstone, for example, which for a hundred years or so resided in Savannah, Missouri, is credited with curing more than fifteen hundred cases of hydrophobia with only a single failure.

Bezoars form in the stomachs of several animals—in horses, mules, whales, dogs, even lizards—but none are so highly prized as those from deer. Modern medicine predictably doesn't think much of their curative powers, probably with good reason, but there still are old people who would apply a madstone to a dog bite before they called a doctor. No sense taking chances.

As yet another internal oddity, none of the cervids except the musk deer have gall bladders, a characteristic they share with many other herbivores. Gall bladders store bile, which is of use in digesting certain fats, and animals that alternate fasting with eating jags need a fair amount of it when they load up with food. The cervids' diets and their habit of eating almost constantly make a reservoir of bile unnecessary.

Which is not to say that deer aren't decidedly bilious at times. It happens in the fall, under the Rutting Moon, when a buck deer goes off a-courting. He's normally a retiring sort, slipping around his favorite haunts shy as a shadow, but in the rutting season, looped to the gills on male hormones, his neck swollen twice its normal size, he's brash, belligerent, and a thoroughly dangerous animal.

Much of his self-image and nearly all his aggression centers on his antlers—not surprising, since antlers also are the major part of what makes deer so interesting and appealing to us.

Unlike horns, which comprise thin layers of a substance originating from the epidermal layer of skin, antlers are solid, bonelike, and grow directly from the animal's skull. Every cervid grows a new set each year.

The process begins in spring, when lengthening hours of daylight stir the pituitary gland into action. The animal's blood carries increased quantities of calcium, protein, phosphorus, and other material to two pedicels on the frontal bone of his skull. There antlers begin to grow, developing from the tips outward like the branches of a tree, forming the shape characteristic of the species.

Antlers grow for several months, all the while covered with a thin layer of skin. This skin is soft, richly supplied with blood vessels, and covered with short, velvety hair. While still "in velvet," as the growth stage is called, the antlers are soft and easily damaged or deformed. They reach maximum size at the end of summer, usually in September, their growth halted by increased levels of testosterone in the buck's system. At that point blood ceases flowing through the velvet sheath, and the membrane begins to dry and wither. The antler itself hardens to the consistency of bone.

The number of tines on his antlers does not reliably tell a buck's age, since available minerals, particularly calcium, play a key role in antler growth. Their shape, however, offers at least some notion of how old he is. Very young bucks, in the first autumn of their lives, may show only small bumps or buttons in front of their ears. A yearling's antlers are always visible, sometimes no more than short spikes, sometimes once-forked. As he ages his antlers continue to grow more branched until each one carries six to ten points—although hormones and diet affect antlers profoundly and at times create bizarre racks that look more like rakes or brushpiles than cervid headgear.

A buck's antlers reach maximum size and mass when he's about eight years old, if he lives that long, and from then on they likely will grow progressively smaller, more irregular, and less symmetrical each year. Owing to a radical imbalance of hormones or to a genetic quirk some bucks never grow antlers and a few does do.

In a typical healthy animal, however, the antlers are in prime condition at the onset of breeding season. At that time he begins to mark off his turf with a series of rubs and scrapes. He creates rubs by thrashing shrubs and small trees with his antlers, shredding bark and wrecking twigs and branches. Bucks once were thought to do this because the dried velvet on their antlers itched; even though such shadowboxing does remove the last shreds of velvet and polishes the antlers, mammalogists now are more inclined to see rubbing as dominance behavior.

A scrape is a scent station, a section of ground perhaps a yard square, which the buck paws vigorously with his front feet.

He perfumes the area with urine and with secretions from the scent glands on the insides of his hind legs at the knee joints. Scrapes serve both to warn other bucks away and to attract does, who may also urinate on them to establish contact with the resident buck.

So long as his testosterone surge continues he's on a rampage of lust and battery, fighting with other bucks to maintain his dominance, mating with as many does as he can, and sleeping little in the meantime. He may lose as much as a quarter of his body weight before the breeding season ends. The fights usually amount to shoving matches, antler to antler, and end without serious injury to either party. Occasionally, when a particularly aggressive buck slips under his opponent's guard, only one of them walks away.

Bucks always initiate breeding, to which the does may or may not respond. Since the does come into and out of estrus over a period of months, as early as September or as late as March, a given buck sometimes has a tough time finding enough willing females to slake his lust, and the more hotheaded of them occasionally will kill an unresponsive doe.

This is extreme behavior, of course, but nearly every sexually mature buck spends the rutting season on a hair-trigger edge of violence. Those who go soppy at the sight of deer placidly browsing in a summer meadow and who natter on about the "big brown eyes" should see the same creature at the height of the rut, when he behaves like something out of a nightmare. With the combination of antlers and sharply pointed forefeet he's extremely well armed and afraid of nothing. More than one old deer biologist can show you some mightily impressive scars and tell you stories to curl your hair. Having had some experience with both, I'd sooner share a phone booth with a rattlesnake than a two-acre enclosure with a rutting buck.

Curiously though, if you saw off his antlers even the most bellicose and sex-mad buck instantly becomes Mr. Milquetoast—or more accurately Dr. Jekyll, reverting to the character he is during the rest of the year. He'll lose them anyway, come winter, when the breeding season ends and his testosterone charge subsides. Depending upon the latitude where they live, bucks shed their antlers sometime between the end of December and mid-April. A decreased hormone level is the primary catalyst, although nutrition seems to play some role as well; the better fed he is the longer a buck tends to carry his antlers. In any case they eventually separate from the skull pedicels and drop off one at a time, within a few hours or a few days of one another.

Even where deer are abundant, finding a cast-off antler by chance is a fairly rare event, rarer still if you're not in the woods at shedding time. Folklore has it that bucks bury their disused headpieces, but what really happens to them is much simpler. For one thing, antler material decomposes more quickly than skeletal bone, and for another, rodents relish its high content of chemical salts. If you do find a shed antler, even a fresh one, it's likely to show myriad toothmarks where mice, wood rats, chipmunks, squirrels, or porcupines have been at it. In areas where the water is low in bone-building minerals deer will chew on their discarded antlers, too.

By that time of year they need all the nutriment they can get, especially in the north where winter locks down with the cruelest grip. When cold weather hits, deer begin drifting toward cedar swamps and conifer thickets—"yarding up," as the process is traditionally called, where they are most sheltered from wind and deep snow. Among northern deer, winter range may be no more than a tenth of the area they occupy during the rest of the year.

Northern white cedar is superlative deer food, but if too many animals crowd into a small thicket or if the winter is a long one, demand outstrips supply. Deer yards frequently become death traps, for the animals refuse to leave them once the snow is more than about two feet deep. Deer in deep snow are sitting ducks for predators, and their insistence upon remaining in the yards may be as much an instinct for avoiding being eaten as it is a search for shelter. That the yard may become utterly devoid of food from ground level to as high as a deer can reach doesn't seem to matter. Starving deer trapped in bare, eaten-out yards have been transported to plentiful food as much as three miles away—and promptly trekked back to go on starving in the same old place.

Deer in the southern ranges don't yard up, but high water can be as deadly as deep snow; even though they're powerful swimmers, the instinct to remain in one place during the gut of winter is stronger still. A northern American winter may kill as many as 3 million deer.

On the surface that may seem a persuasive argument for abolishing deer hunting, but like most superficial reasoning it hardly accommodates reality. Now that we've eliminated most natural predators from all but a minuscule portion of deer range and similarly have reduced wintering habitat to a minimum, sending more deer through the gate of winter simply increases the number that will not come out on the other side and in addition inflicts further hardship upon those that do survive. A severely winter-stressed, pregnant whitetail doe may resorb her fetus or, if she does give birth, may find herself unable to produce enough milk that the young can survive. Weak, malnourished animals of any kind do not reproduce well nor do the young born of such parents, thus passing the deterioration of a species' well-being on through succeeding generations.

When properly managed according to the realities of nature—a process in which hunting plays a vital role—deer demonstrate a remarkable ability to survive all manner of natural disaster, and the reason for this is most apparent in the spring.

Gestation among the Cervidae is a relatively long-term affair, ranging from about 160 days for musk deer to as long as ten months for roe deer, which are the only cervids in which delayed implantation of the fertilized egg is known to occur. Whitetail does carry their young for about 200 days, so a doe mated in October will give birth in April or May.

How many fawns she bears depends upon her age; if she's about eighteen months old and in her first pregnancy there likely will be only one. Older does usually bear twins, sometimes triplets, and in rare instances quadruplets, although one of the four almost invariably dies shortly after birth. Of every hundred fawns born there will be one or two more males than females.

Regardless of sex, the newborns are about eighteen inches long and weigh from four to seven pounds each. They enter the world with their eyes open and dressed in a coat of russet hair patterned with rows of white dots and flecks on the back and sides—all of which gets a thorough licking from the doe immediately after birth. Within a few minutes the fawns are able to stand on their tottery legs and can nurse while standing by the time they're a half-day old.

In the early weeks they nurse briefly every two or three hours, rapidly gaining weight and strength from their mothers' rich, highly concentrated milk. Otherwise they keep a low profile, lying curled up or stretched flat in a patch of sun-dappled shade or amid low-growing vegetation. Natural camouflage and instinctive immobility are their primary defenses, apparently helped early on by a remarkable lack of body scent. Precisely how little odor very young deer give off has probably been exaggerated by folklore, but even sharp-nosed dogs can pass within a few feet of a motionless week-old fawn and not know it's there.

The scentless period lasts only about a month, until the fawns' scent glands begin to function, but by then they have less need for anonymity. They've begun following the doe on her daily round of feeding and have themselves begun to eat solid food. Some will continue nursing until they're about six months old, and even after they're fully weaned they remain with their mothers until they reach breeding age.

By midsummer adult whitetails are wearing their reddish warm-weather pelage, which they will molt in September for a thicker coat of grayish hair. By late summer the fawns have changed their spotted natal coloration for adult garb and are essentially smaller copies of their mothers. Their fathers haven't taken the slightest interest in matters paternal and have in fact ignored fawns and does alike since the end of the breeding season, preferring to hang out by themselves or with a few cronies. Their stag parties—you should pardon the pun—will continue until the rutting moon comes again.

That bucks assume no parental responsibility seems to make no difference in the survival rate among the young. By the time they're six weeks old the fawns are strong and nimble enough to elude much of the trouble they're likely to get into. Does moreover are attentive and protective parents, willing to brook no guff from foxes, dogs, coyotes, and other predators. A couple of summers ago I watched a doe drive a full-grown coyote away from her fawn literally in my front yard, feinting and slashing with her forefeet and eventually sending him off in full rout. I couldn't blame him for choosing discretion; angry as she was, I wouldn't have messed with her, either.

In some areas wolves, mountain lions, bobcats, and lynxes account for a share of deer, both young and adult, but the extent of the predation generally is negligible. Over most of the range domestic dogs, whether freelancing farm and suburban pets or truly feral specimens, are by far the most serious predators of deer, and the automobile is the prime cause of accidental death. Still, their high reproductive rate coupled with the fact that does fairly often live ten or fifteen years in the wild more than offsets the annual mortality.

Man historically has been the deer's greatest enemy and more recently the author of its salvation. To the North American Indians living east of the Great Plains, white-tailed deer were what the buffalo represented to the western tribes. Virtually every part of the animal from its brains to its sinews served some useful purpose. At certain excavated Indian sites in Missouri, known to have been occupied before 1600, about 90 percent of animal bones found belonged to deer.

◆

Deer also were basic to the white man's survival in North America. Wary, fleet of foot, and possessed of incredibly keen senses of smell, sight, and hearing, deer are wonderfully well suited to survive, but they were scarcely a match for the influx of man. Hunting pressure was so intense that even colonial governments recognized the effects. Massachusetts, for example, passed its first law regulating the harvest of deer in 1698. Other colonies followed suit but with little or no appreciable effect. As a result of habitat destruction and year-round hunting both for subsistence and for the wild-game markets, deer east of the Mississippi were almost a memory by 1890.

The situation wasn't much better anywhere else in the country. Before Europeans arrived in the New World as many as 40 million white-tailed deer may have inhabited 2 million square miles of range. By 1900 perhaps 500,000 remained. In 1925 the deer population in all of Missouri was believed to be no more than 395 animals; only 3 were known to exist north of the Missouri River.

The renaissance, ushered in by the first great age of modern conservation and wildlife management, was even more dramatic. Over the next thirty years intensive efforts in stocking and redistribution, combined with restrictive—and strictly enforced—hunting regulations, met with astonishing success on a national scale. What happened in Missouri is a good example of what happened nearly everywhere. By 1935 the Missouri deer herd had grown to an estimated 2,240 animals; in 1959, for the first time in the twentieth century, there were deer enough to support a hunting season in all 114 counties; and in 1973, 2,529 deer were killed by automobiles—more than existed in the entire state only forty years earlier.

All those deer were a mixed blessing. In Missouri as elsewhere, restricting the hunting harvest to antlered bucks had been a key element in the population rebound. Their polygamous ways make a substantial portion of the annual buck population irrelevant to the species' reproductive success, and confining the hunters' kills to those surplus animals worked beautifully—better in fact than anyone guessed. As early as the 1920s deer in the East and upper Midwest were outgrowing their winter food supplies and state after state began to witness massive die-offs—Pennsylvania in the 1920s, Wisconsin in the 1930s, Michigan in 1950–51, and almost every state during the particularly harsh winter of 1955–56, when more than 2 million deer starved and died in the United States.

Game managers recognized the overpopulation problem early on and realized that the herds needed more thinning than bucks-only hunting could provide. So for the next forty years they found themselves in the ironic position of arguing in favor of shooting does and yearlings—of trying, in other words, to undo some of what they already had done so well.

It was an uphill battle against resistance from public sentiment, legislatures, and even tradition-bound sportsmen whose understanding of deer biology amounted to little more than folktales, and who never saw the pitiful remains that clogged deer yards all across the north every spring. An uphill struggle but ultimately a successful one, for overly numerous deer are brutally hard on habitat, and farmers, foresters, and other land managers began to see that the biologists were right. State legislators and deer hunters gradually began listening to the combined voices of those who knew the facts, and any-deer hunting seasons now exist in states that formerly allowed only bucks as legal game.

And the dire predictions of those who prophesied that shooting does would wipe out deer populations altogether could not have been more wrong. On the contrary, there are more white-tailed deer in America today than at any time in this century. Around the Great Lakes deer may in fact be more numerous now than even in aboriginal times. Here in Missouri, where they were all but exterminated less than seventy years ago, whitetails are plentiful enough to be a genuine nuisance in some places; despite an annual harvest of better than 100,000 the population probably has not yet topped out.

◆

No other animal of its size is so adept at living close to man and remaining virtually unnoticed all the while. I live in the most popular resort area in the Midwest, three miles from one village and a mile from another, and on any evening you can see six or eight deer grazing in the pasture in front of my house. Granted, I'm down here in the woods, but I've often seen deer just at daylight cropping the twenty-foot strip of grass between the multiple lanes of U.S. Route 54 and the parking lot of our local factory merchants mall, which covers more ground than a lot of farms did when I was a kid.

Actually, as a boy I could see deer on almost any evening simply by looking west from the barnyard at my aunt's house, across the valley to where the grainfields met the woods. And as carefully as I watched I never saw them arrive. They simply appeared, spirits of the woods come to make the sundown somehow more complete. The feeling I got then hasn't left me now, especially when I realize that the deer I see are scarcely a fraction of those unseen. And I'll always remember the one time when I saw them all.

We'd been a week in the wilderness, some friends and I, on a spring backpack trip into the foothills around Laramie Peak in Wyoming. Driving home through western Nebraska, late afternoon found us on the portion of Interstate 80 that runs parallel to the Platte River, perhaps a quarter-mile distant. It was an unusually wet spring, with the riverbottoms flooded, and the high water had pushed the deer from the willow brakes.

I imagine you could drive that route on any mild evening and see a few deer in the wheatfields and meadows between the highway and the river, but that day it was a truly awesome sight. Someone commented on the first ones, then pointed out others farther back near the riverbottom thickets, and then still others up ahead. The more we looked the more deer we could see, quietly grazing in the winter wheat—scores of them, hundreds, dotting the lush green for miles.

◆

Something magical came over the farm country and over us as well, an understanding that we were being treated to a rare and lovely view of something that usually is hidden beneath the bland, contrived veneer of civilized landscape. For a long while no one spoke. We simply watched the deer through the gathering dusk, emissaries of a world that penetrated time and space, a world unrolling smoothly back toward the beginning.

# 15

# Lord of the High Country

The first wild elk I ever saw seemed more ghost than animal. They still do, almost twenty years later.

That spring a couple of my colleagues and I decided to celebrate the end of our teaching year with a week's backpack trip through the wilderness around Laramie Peak in southeastern Wyoming. Monte, who had taken his doctorate at the university in Laramie, spoke highly of the country. Trundling our VW Bug across the sere, roadless, shortgrass plains toward the distant scarp, I began to see why. Presently, standing on rancher Art Fawcett's back porch and looking up at the beetling, stony shoulder of the ridge they call Valhalla, I knew for sure. Westerners often take the sheer wildness of their countrysides for granted; flatlanders on holiday do not. At 10,272 feet, Laramie Peak is no great shakes as mountains go, nor does it offer the stark spectacle that younger, barer ranges do. It is more intimate country, the ridges thickly timbered above open, grassy valleys and parks, laced with streams that rush cold and brisk toward the plains below.

We left the car at the ranch, shrugged into our packframes and headed in-country to set up our first camp at Lost Creek, which the map showed to be about three miles off. Maps, however, measure distance as the crow flies, not as the hiker walks, and after crossing the first ridges I began to wish I'd put three months into pre-trip conditioning instead of only one. During a rest break, after gasping in enough oxygen that I felt like wasting some breath on talk, I remarked to Monte that the local mule deer must be both plentiful and remarkably healthy. Nothing decomposes very rapidly in the cool, dry high-country atmosphere, and the ground we'd covered was strewn nearly everywhere with deer pellets.

Not so, Monte said. Elk pellets. An hour later, crossing the spine of a ridge, we saw the first two cows at the crest of the next ridge over, two shapes that simply faded into the pines like whiffs of smoke.

"That's how you see them most often," Monte said. "They can hear you coming, so they wait till they locate you by smell or sight and then just disappear."

That seemed an intriguing challenge, so I veered off a hundred yards and tried some impromptu stalking. I toiled up each slope as quietly as I could, using trees and bushes for cover as I neared the tops, peering ahead with what I imagined to be all the stealth of a Cheyenne hunter. And often as not the elk were there—always providing a tantalizing glimpse and a vanishing act, until I finally decided that sneaking up on an elk is a job for a better sneaker than I.

Likely it has always been thus, for I doubt the nature of elk has changed much in the million-odd years since its ancestors crossed the Bering land bridge from Asia to North America. What has changed in historic times is the territory it occupies. For the most part we think of elk as western animals, and for the most part they are. Originally, though, they ranged virtually throughout the continent, from northern British Columbia to New York state, from central Georgia to the California coast, sharing the eastern forests with white-tailed deer and the prairies with bison and pronghorn. Both Jacques Cartier in 1535 and LaSalle in 1682 remarked on the great numbers of elk in the Northeast and upper Midwest. In his splendid 1966 essay on elk, John Madson says they were so numerous in the East that early settlers used their trails as roads through the wilderness.

British colonists found the animals familiar enough, since American elk and Scottish red deer are the same species, but the name they chose is a miscue. Technically speaking, *elk* is the European name for the animal we call *moose,* while the animal we call *elk* actually is a species of deer, related to eight other species found all around the Northern Hemisphere.

Everyone it seems has known the animal by a different name. It was *eh-kahg-tchick-kah* to the Sioux, *wapiti* to the Shawnee, Canadian stag to Canadians, and gray moose to some early Yankees. More recently taxonomists have had their own go at it, first insisting that the American animal was a separate species altogether, *Cervus canadensis,* replete with a half-dozen subspecies. Now they're content to view elk and red deer as *Cervus elaphus* and prefer the old Shawnee word as a common name. But then there are those who go right on using *elk,* figuring one name is as good as another. The animals themselves, of course, couldn't care less.

For our purposes, let's call them elk and be done with it. They are magnificent creatures by any name.

Among the deer family only the North American moose is larger. Exact size varies according to season, living conditions, and geographic range—and so to an extent does coloration, which seems to have been the main reason why they once were classified in the various subspecies. At any rate, the average elk stands nearly five feet at the shoulder and measures about seven feet from nose to tail. A full-grown bull averages about 700 pounds live weight, perhaps half a ton when laden with late-summer fat.

Among the overall population, geography seems to influence size more than any other factor. The little tule elk of California, which seldom weigh more than 400 pounds, are the smallest. Roosevelt elk of the Pacific Northwest are the largest; big bulls often run 1,000 to 1,200 pounds, and specimens of 1,500 pounds or more are not unheard-of. Even these behemoths are second in size to moose, but not by much.

Although big-bodied, elk are small-headed. You notice this among cows and even bulls in spring and early summer. But you can look for a long time at a full-grown bull in fall and winter and never notice anything but his magnificent antlers. In sheer mass elk antlers cannot match the broadly palmated headgear of a bull moose, but elk antlers are big enough. The main beams, thick as a big man's wrist, grow fifty inches long or better. Especially large ones have measured sixty inches and more. Even an average set might weigh somewhere between forty and fifty pounds. As do other members of the family, bull elk shed and regrow new antlers every year, beginning when they're about a year old. The first set will be little more than ten- to twenty-inch spikes. Next year, and perhaps even the year after, he'll show four or five points, but after that he'll typically wear a rack with six points on each side and be therefore known as a royal stag. Some develop a couple of additional points and are then called imperial bulls.

Each of the six points in a standard set of elk antlers has a name, handed down from the sporting traditions of medieval Europe. The first branch, nearest the base of the antler, is the brow tine, and the next one up is called the bay or bez tine. Both of these point forward, and between the two antlers the four tines together make up a serious set of weapons, nowadays collectively called the lifters, formerly dog-killers or war tines.

Above these, the third tine is known as the trey or trez, and higher yet, where the antler reaches its outermost curve, is the royal or dagger-point. The last two tines, formed by a fork at the tip of the main beam, are called the sur-royals.

◆

Elk headwear, like that of deer, is made of bonelike material and grows from pedicels on the skull. And it grows at an astonishing rate. Elk typically shed last year's rack in March; new antler buds begin forming in May and are fully developed by the end of September, having grown at the rate of nearly ten pounds per month.

Once the antlers reach full size, the blood-rich, velvety membrane covering them dries, cracks, and begins to peel off. The animal helps the process along by rubbing his rack against trees and thrashing at bushes—behavior that is some indication of things to come since bull elk, like all deer, grow an attitude along with their antlers.

Although elk are decidedly gregarious animals—and in being so are different from most other deer—mature bulls generally don't hang out with the herds in late spring and summer. Instead they spend the warm months alone or with a few cronies and leave the business of herd-living to the cows, calves, and yearlings. (In social structure an elk herd is in fact a matriarchy, and despite the bulls' lordly appearance, the true leader invariably is the oldest cow.) In the fall, however, their antlers fully grown and their systems brimming with sex hormones, the bulls drift back to the herd, primed for procreation and cocked for a fight.

At first they do battle with the local vegetation, slashing and shattering sapling trees, bushes, and decaying logs—behavior designed to impress both the cows and other bulls, to assemble the former into a harem and warn the latter of what will happen should they interfere. They further announce their intentions by bugling, a sound part bellow, part grunt, part whistle, and altogether one of the most thrilling vocal displays in nature. It rings and echoes among the high-country peaks like the winding of Gabriel's horn, a sound that is, as Madson puts it, "the distillation of all mountain wilderness."

They stake out their harems by intimidation and force, driving together as many cows as possible and driving off the immature bulls still with their mothers. The youngest bulls of breeding age play the spoiler's role, hanging around the edges, trying their best to split off a few cows, and generally making nuisances of themselves until the herd bull one way or another gives them the gospel.

Sometimes, though, the interloper is just as big, just as mature, just as bellicose as the harem-master himself, and the subsequent battles are as spectacular as you'd expect between animals that might weigh a half-ton each. They roar and charge, smashing their antlers together with incredible violence, shoving and lunging with all the enormous strength of their shoulders and necks. Naturalist William Graf wrote in the 1950s of finding a dead bull with both antlers torn off from the impact of a head-on charge against another bull. Plates of skull bone were still attached to the antlers' pedicels. Such fights to the death inspire a full measure of awe, but they are uncommon. Herd bulls usually are able to keep matters in hand through a combination of bluff and minor skirmishing.

Which is well, because fighting isn't their main interest anyway. Most animals are aggressive breeders during the appropriate season, but a bull elk in full rut is a stud of monumental appetite. To say that he's highly polygamous and promiscuous understates the case. Wild elk have been known to attempt mating with dairy cattle and female horses, and Madson describes one particularly enthusiastic Oregon elk who beat up a Hereford bull in order to take possession of a domestic cow herd.

By the time the rutting season ends, usually in October, the dominant bulls are worn to a frazzle and the cows have begun the eight-and-a-half-month job of gestating the next generation. Cows and younger animals re-form their herds, the bulls catch some much-needed rest, and all wait for the snowstorms that will trigger a migration to the wintering grounds at lower altitude.

Although essentially grazers that prefer various sorts of grasses to other foods, elk aren't averse to browsing twigs, buds, evergreen needles, leaves, or almost anything else that grows. Such eclectic taste stands them in good stead, especially during winter. Snow poses few problems since elk are big enough to paw through a considerable depth to get at the grass underneath. If their wintering range is overgrazed they switch to browse, reaching as high as nine feet above the ground.

A varied diet notwithstanding, winter brutalizes elk just as much as it does other animals. Starvation is a major cause of death among elk—that and the diseases they contract in their winter-weakened condition. The most important of these, however, is not brought on by weakness but rather, in a roundabout way, by their food itself. Sharp seeds, twigs, splinters, grass awns, rushes, and hay stems pierce and lacerate the lining of their mouths and expose them to invasion by the bacterium *Actinomyces necrophorus* and a gruesome death from necrotic stomatitis. The infection creates severe lesions in their mouths, nasal passages, tracheas, and lungs and may even extend through their digestive systems. The actual death, however, comes as blood poisoning from a toxin the bacteria create as they multiply.

Good evidence suggests that necrotic stomatitis is most prevalent on overpopulated range where the animals are artificially fed on baled hay during winter—but fossilized bones dating back to the Pliocene also show that *A. necrophorus* has plagued elk for at least 10 million years. The disease may ravage certain localized populations, but it clearly has not greatly affected the species as a whole.

Nor for that matter has any other ailment or parasite, and you can see evidence of that in the fact that elk sustain a low birthrate. One calf per breeding cow per year is the rule, twins the rare exception. Born in late May or early June, typically in open sagebrush with timberland nearby, a newborn elk is a leggy, wobbly little specimen of perhaps thirty-five pounds; it spends its first couple of days lying around between attempts at taking a few staggering steps. By about five days it's a good enough runner to stay out of at least a portion of harm's way. Its main defense, though, comes in the form of an extremely attentive, affectionate mother and a cadre of equally alert aunts, sisters, and female cousins all large enough to stomp a dog, bobcat, or coyote into hairy marmalade. Wolves, bears, cougars, and golden eagles all take a certain toll on elk, but as with diseases and pests no predator poses a serious threat to elk as a species.

With one exception. As I said earlier elk ranged over most of North America in presettlement days, and the arch-predator, man, is the sole reason why they no longer do.

When European man arrived, the elk population of the New World may have been as high as 10 million. By 1919 the total in the United States was down to an estimated 70,000. Just as they had for the Indians, elk provided American colonists and settlers with a bounty of food, hides, and material for tools. I can tell you for sure that a roast from even a middling elk is better eating than the best venison haunch you ever tasted, and leather tanned from elk hide can be soft as old cotton or tough as tin, depending on what you need it for. There is little call nowadays for sinew, bone awls or fishhooks, but one elk skeleton and a set of antlers can supply enough raw material to make an abundance of such items.

Even by the turn of the nineteenth century the great eastern herds were nearly gone. Lewis and Clark, who kept meticulous journals from the moment their boats left St. Louis on May 14, 1804, do not mention elk until the party reached western Missouri—although the animals hung on in some parts of the state until almost 1900 and are namesakes of two Missouri towns.

Some eastern elk even outlasted Lewis and Clark, though not by much. The last one in Indiana was shot in 1818, the last in Ohio in 1838. Elk remained in the Genesee Valley of New York until 1847, in Tennessee until 1849, in Iowa and Virginia until 1857. By 1881 they were gone from Nebraska and Oklahoma, from North Dakota by 1890, from Minnesota by 1900. Western populations fared much the same during the first two decades of this century.

◆

At bottom the slaughter had most to do with subsistence living, with a thriving commercial trade in wild game, and to a lesser extent with teeth.

Somewhere in their evolutionary past elk had fangs. A few species of Asian and South American deer still do—three-inch tusks that grow from their upper jaws. In elk these teeth are vestigial and serve no known purpose since they don't occlude with any corresponding teeth in the lower jaw. Nonetheless they seem to wear at the same rate as an elk's truly functional teeth, and they're larger in bulls than in cows.

The Indians thought elk tusks good medicine, prized them as amulets and charms, wore them as jewelry and as decoration for clothing, and passed them on from one generation to another. The white man agreed, and around the turn of the century elk teeth capped in gold were popular adornments on many an urban gentleman's watch chain—especially among members of the Benevolent and Protective Order of Elks, the fraternal lodge that still honors the elk as its symbol.

Eastern herds already were seriously depleted or extirpated, and the elk-tooth fad helped hasten the western population's decline—5,000 killed around Yellowstone during the winter of 1915 alone, solely for their teeth. Commercial tusk-hunters made similar inroads all over the West. It was outright poaching, of course, because game laws already were in place over much of the country and early restocking programs were by then underway in some areas. The laws, however, had no more teeth than a dead elk, and the illegal killing continued until the Elks Lodge officially banned its members from wearing the tusks and the game wardens received better political and legal support.

For American elk the road back from the brink of oblivion was remarkably short but not without some perilous turns. National parks, especially Yellowstone, were the key source of animals for live-trapping and relocation, and elk are hardy enough to tolerate the stresses of being caught and moved. Regrowth from logging, forest fires, and from great tracts of conifer woods killed by an Engelmann spruce beetle epidemic provided abundant food. Wherever habitat conditions were suitable, elk prospered.

And prospered and prospered. A healthy elk population increases its numbers by about 20 percent each year, and a herd that isn't hunted can double in four years. In Yellowstone overpopulation was evident even before the turn of the century, and by 1920 thousands of elk had starved to death in the northern park and in the Jackson Hole area. Similarly massive die-offs occurred periodically in Washington and Idaho, and in other parts of the West.

Once again, the problem was man. Except in the parks, which are a somewhat artificial situation, elk are not nearly as amenable to farmland living as white-tailed deer, and by the early twentieth century man had usurped the greater part of elk range, not just in the East but everywhere. What remained would support the animals in reasonable numbers, but all the elk knew to do was follow nature's arithmetic—keep multiplying and thereby divide a dwindling food supply among an ever-greater number of mouths. When that happens the entire population suffers, the range itself suffers, and in the end nature goes on a brutal rampage of subtraction.

But elk don't give up without a fight. Rather than starve peaceably in the wilderness as deer are wont to do, elk invade farm and ranch land, eating everything from pastures and row crops to orchards and haystacks. No other big animal in America competes more directly with domestic livestock nor more effectively plunders cash crops.

Modern, scientific wildlife management is a fabric knit of two main threads: Maintain good habitat and wildlife populations will respond; control wildlife populations and the habitat will remain healthy. Hunting, with seasons and limits properly regulated and strictly enforced, confers far more benefit to game species than any amount of winter feeding, legislation, or hand-wringing. By the early 1950s the states holding the greatest number of elk finally caught on and the animals have for the most part been intelligently managed ever since.

Even some eastern states—Virginia, Pennsylvania, and New York, for instance—have a few elk once again, and Michigan's Pigeon River Country supports the largest herd east of the Mississippi, 1,000 animals maintained and kept healthy by a limited annual harvest. I hunted grouse and woodcock in the Michigan range last fall, and while I didn't see any elk I saw plenty of evidence that they're alive and well and possess highly functional digestive systems.

Walking the regenerating clearcuts amid scattered piles of elk pippies took me once again to Wyoming and the Laramie Peak country. I finally did get a good look at the elk on that trip, exploring alone while my friends made a day-hike to the top of the peak. I was sitting on a rock near the foot of a slope watching a pair of ouzels, those curious little dipper birds, feeding underwater in the stream just below. After a while I happened to look up and there at the edge of the trees on the far side of the valley stood three cows, grazing in the lush spring grass. With the wind blowing down-valley between us they never caught my scent, and I was settled comfortably against the rocks to begin with so I had no reason to move. I sat and watched for nearly an hour while they placidly cropped away, drifting slowing up the valley. I stayed a while even after they moseyed out of sight around a little neck of timber, wishing I could be in that same place about four months later to hear the valley echo and ring with the herd bull's fanfare of bugles.

And then I got up, stretched, and headed slowly downstream toward camp, thinking about how far above the flatlands I had really come.

# IV

# Splendor and Spectacle

Kenneth Wind

# 16

## Bird of the Gods

On June 20, 1782, the Congress of the Confederation adopted a design for the Great Seal of the United States in which the image of a bald eagle is the central motif. As I mentioned in an earlier chapter, by 1784 Benjamin Franklin was having second thoughts about whether he and his colleagues had chosen the proper bird. "For my part," he wrote to his daughter Sarah Bache, "I wish the Bald Eagle had not been chosen as the representative of our country. He is a bird of bad moral character; he does not get his living honestly...Besides, he is a rank coward...The turkey is a much more respectable bird."

Though his compatriots usually took Franklin's views seriously, such misgivings about the eagle would have earned little sympathy. Virtually throughout the Northern Hemisphere eagles have been symbols of strength and majesty ever since Paleolithic man rendered their likeness on the walls of caves.

To the ancient dynasties of China, to the Assyrians, Babylonians, Greeks, and Romans eagles were emblems of courage, military might, and the majesty of empire. To the Greeks the eagle was the bird of Zeus, to the Romans the bird of Jove. The Roman legions marched under the silver and gold emblem of eagles and believed that an eagle delivered a dead warrior's soul from flesh to immortality. Eagles adorned the banners of the Holy Roman Empire, of Imperial Russia, of Napoleon, of Austrian and Byzantine emperors. To some North American Indians the eagle symbolizes ancestral immortality. Others saw the great war eagle as an emblem of courage and triumph.

Among falconers since ages before Christ only an emperor or king could fly an eagle. The berkut, the Asian species of golden eagle, was prized as the greatest hunter of all, flown by the emperors of China, of Japan, by Alexander the Great, by the Russian czars, by Genghis Khan, Saladin, and Richard I of England. Even now, according to a tribal khan in the Kirghiz Republic, a trained berkut is worth "a dozen young and lively wives and more than twenty Bactrian camels."

Eagles have long been supposed to possess even stranger powers. In *Naturalis Historia*, completed in 77 A.D., Pliny the Elder says that eagles are immune to death by lightning and that they can rejuvenate themselves by diving into a well. Pliny also reports that stones gathered near an eagle's nest confer the power of invisibility upon the man who takes them. The Greeks believed that burning the right wing of an eagle in their vineyards protected the vines from hailstones.

Some medieval physicians ascribed certain medicinal properties to the marrow of eagles' bones—if nothing else good evidence that medicine was not yet an empirical science, as virtually no birds' bones contain marrow.

So in the face of almost five thousand years of recorded opinion to the contrary, what was Ben Franklin's beef with eagles? Actually his objection was to the bald eagle, a bird only a handful of Europeans and even fewer Asians had ever seen, and setting aside his comments about moral character and cowardice, Franklin was essentially right.

Eagles belong to the order Falconiformes and to the family Accipitridae, the birds of prey, of which there are ten subfamilies comprising sixty-five genera and more than two hundred species worldwide. Eagles, buzzards, and the true hawks make up either the subfamily Accipitrinae or the subfamily Buteoninae, depending upon which group of taxonomists you choose to side with. There are about fifty-five species of eagles worldwide, and they inhabit every continent except Antarctica. Only two species breed in North America, however, and at the end of the eighteenth century Ben Franklin apparently was one of only a few Americans who recognized the more important differences between them.

The colonists probably saw golden and bald eagles in about equal numbers. They're about the same size and are hard to overlook as they can measure nearly a yard from beak to tail-tip, weigh as much as fourteen pounds, and have a wingspan of more than seven feet.

Being Europeans, the colonists' image of eagles as fierce, proud, deadly hunters derived from the golden eagle, which lives all across the Northern Hemisphere. The notion had plenty of basis in fact, since golden eagles are splendid predators and sometimes kill prey nearly as large as themselves. Royal falconers, in fact, flew berkuts at wolves, deer, bear, and in pairs even at leopards and tigers. It was conditioned behavior, of course, because wild eagles in search of food simply don't attack prey that large, but that the birds were willing to do so at all bespeaks a ferocious temperament (no doubt influenced by the old falconer's habit, now long out of practice, of maintaining his birds at the edge of starvation).

The bald eagle, on the other hand, lives only in North America, and while its plumage is clearly different from the golden eagle's, the European immigrants must have assumed that its personality wasn't. Preoccupied with the problems of surviving in the New World wilderness, few of them cared at any rate. But Franklin, ever the astute observer if not an enthusiastic naturalist, must have noticed that bald eagles seldom kill any sort of prey and when they do it's usually a small seabird or a fish snatched delicately from the water in a low swoop. He may have seen them along the seacoast and the rivers feeding on dead fish and assorted carrion, pirating the catches of ospreys and other fishing birds—generally behaving like scavenging layabouts who don't come by their living through honest toil.

Still the Continental Congress could hardly have chosen a nobler-looking bird as the emblem of a nation. With its brownish, charcoal-gray body plumage, snowy-white head and tail, piercing eyes, hooked, bright-yellow beak, and lordly bearing a bald eagle is as handsomely patrician as any senator ever hoped to look.

In appearance alone the golden eagle is less striking. Except for the bronzy nape feathers that give the bird its common name, its plumage is a nondescript dark brown with patches of white on the wings and tail. Its beak, smaller than a bald eagle's, is a bluish bone color with a dark tip. But the golden eagle's slightly seedy dress and the bald eagle's repulsive tastes notwithstanding, there is much to admire in both of them, for nature has provided eagles with some extraordinary gifts.

It's hard to look away from an eagle's stare, from the infinite wildness in those brilliant, sulphur-colored eyes, from the intense gaze that seems to leave not even thoughts unseen. Eagle eyes are large, in some species larger even than ours. They're fixed firmly in their sockets, so an eagle must move its head to look around. An alert bird does so continually. Its neck is extremely flexible, capable of such compound curves that an eagle can turn its head upside down like an owl.

Eagle eyes also contain pectens—pleated, cone-shaped structures made up almost entirely of blood vessels and located on the retinas near the optic nerves. All bird eyes have them and their actual function is not clear, but since pectens are largest in the diurnal birds of prey, they may supply additional oxygen and nutriment to the retinas, which have no blood vessels of their own.

Eagle eyes look like high-powered optics and indeed they are. The retinas contain an abundance of cone cells, which promote visual acuity and color perception. Some areas of the retina in fact comprise nothing but cone cells; rod cells, which are sensitive to low levels of light, are less useful to birds active only in daytime. In addition to the cone patches, the retinas of eagles and other diurnal birds of prey have small, curved pits, called foveas, where cone cells and nerve endings are even more densely packed. There are two foveas in each eye, one directed forward, the other sideways. (Vultures, which don't actively catch their food, have only one fovea, directed sideways, in each eye.)

Unlike ourselves, eagles possess both binocular and sideways fields of vision. The binocular field is relatively narrow and has a blind sector immediately in front, which would seem a handicap to a bird that often dives at high speed to catch prey on the ground. Eagles compensate with minute head movements, however, and their eyes can quickly and precisely adjust focus and depth perception.

The resolving power of eagle eyes may not be quite what legend and folklore imply, but it's considerably better than ours. Thanks to the patches of cone cells an eagle's normal visual acuity is about four times greater than man's; at the foveas it's about four times greater yet, which in effect gives eagles something like zoom-lens vision. Using his own visual acuity as a rough guide and applying simple arithmetic, British ornithologist Leslie Brown suggests that with its basic vision an eagle can identify an object the size of a rabbit at a distance of 1,580 yards, or about .9-mile. By moving its head slightly to focus the image at its foveas, the bird can see the same object clearly at 3,160 yards or nearly 1.8 miles away.

Since vision is an eagle's most important sense, its eyes are well protected. A narrow shelf of bone protrudes above the eye socket and serves the same purpose as the bony arch above our own eyes. (The supraorbital process, as it's called, is what gives the eagle its fierce, frownlike stare.) Like many other animals, eagles have a third eyelid, a translucent nictitating membrane attached at the front, which cleans and protects the cornea while still allowing some vision. Eagles draw them across their eyes frequently and nearly always do so when struggling with prey or when presenting food to their sharp-beaked young.

After its eyes, an eagle's feet are its primary tools. All species that kill prey do so with their feet, assisting to a small extent with their beaks. Each foot has four toes, three forward and one behind, all armed with long, sharp, curved talons. Among the prey-killing species the hind-toe talon is particularly daggerlike. Among all the species the feet are highly specialized in size and shape, adapted to the nature and size of the food each bird prefers.

Bald eagles and their fish-eating kin have rough spicules on their toes, which give them a firmer grasp on slippery prey. Golden eagles don't need spicules but instead generally have a more powerful grip and deadlier hind talon. Scutes—tough scale-like plates—armor the feet and lower legs of all raptors, protecting against abrasions, bites from prey, and other injuries.

In all thirty species of the group usually called the true or booted eagles, feathers grow right down the legs to the toes. Why is not clear. It isn't for insulation, since some species of booted eagles live in the tropics. In any case, the golden eagle's baggy-looking knickers do not help promote a regal appearance. None of the fish-eating eagles have feathered lower legs, and that does serve a purpose; water-soaked feathers would only increase the bird's weight and make carrying off a sizeable fish all the more difficult.

Few sights in nature are more impressive than a soaring eagle, although not all species are equally agile on the wing. As with their feet, eagles' flight equipment and aerobatic ability vary according to their livelihood. Because lift is more useful than maneuverability, fish eagles generally have broader wings and shorter tails than the prey-killing species. But even if bald eagles are a bit clumsier than goldens, both are expert gliders, able to plane across wind currents and ride thermals with endless grace.

Soaring is an important survival tool. As do most species, both bald and golden eagles hunt from perches—dead trees, mountain ledges, any vantage point from which the bird can scan for food with its splendid eyes. If a good perch isn't available they soar, because soaring amounts to an aerial perch.

But if you've ever tried to hand-hold powerful binoculars or a telescope or a telephoto lens you can imagine the difficulty of trying to sharply pinpoint some small, distant object while being buffeted by swift, changing air currents. Having wings that flex both at the shoulder and the wrist is an advantage for all birds, but eagles and some other soaring birds have evolved specialized feathers for maximum stability.

Six or seven primary feathers at the tip of an eagle's wing are emarginate—that is, the feather vane is narrowed at the tip. When the wings are extended these feather tips separate like fingers. Eagles in flight control them with astonishing delicacy, constantly adjusting the spaces between them in response to changes in the air, holding the body and head rock-steady even in a gale-force wind.

What the bird is looking for and its reaction at finding it depends upon the species. The golden eagle is a magnificent hunter, usually taking its prey at the end of a long, slanting dive—not quite as impressive as a peregrine's bulletlike plunge but impressive enough. North American goldens feed primarily on rabbits, marmots, and animals of similar size. Carrion is an important source of winter food. They catch most of their prey on the ground but sometimes take smaller birds on the wing. Given the opportunity a golden will kill an occasional lamb, domestic or wild, a mountain goat kid, a fawn, or a newborn calf, but their impact as predators of game animals and livestock alike is vastly overrated. Folktales of eagles snatching human babies are just that.

Bald eagles prefer fish to anything else, though they also feed on seabirds and waterfowl they sometimes kill during the molt, when ducks and geese are flightless. During the salmon runs in Alaska bald eagles gather in huge numbers to gorge themselves on the dead and dying spawned-out fish.

All told, the facts do not justify the eagle's reputation as a rapacious killer. The North American birds require only about eight ounces of food per day in summer and about nine ounces in winter. They do not spend a great deal of time hunting. On still, cloudy days they prefer loafing. Fine weather and some wind seems to stimulate soaring and airborne playfulness.

Courtship display prompts the most elaborate aerobatics. All species perform some nuptial display—calling from perches; soaring, sweeping, undulating through the air (by one or both sexes); rolling and foot-touching in flight. The most famous display of all is the spectacular, rarely seen maneuver in which male and female lock their talons in midflight to whirl and spin downward from a great height, nearly reaching ground or water level before they separate. Only a few species do this, most of them sea and fish eagles.

Walt Whitman seems to have recognized it as a courtship display more than a hundred years ago, though ornithologists have only recently confirmed the fact. Whitman's poem "The Dalliance of the Eagles," written in 1880 from an account by American naturalist John Burroughs, is as good a description of the event as any.

> Skirting the river road, (my forenoon walk, my rest)
> Skyward in air a sudden muffled sound, the dalliance of the eagles,
> The rushing amorous contact high in space together,
> The clinching interlocking claws, a living, fierce, gyrating wheel,
> Four beating wings, two beaks, a swirling mass tight grappling,
> In tumbling turning clustering loops, straight downward falling,
> Till o'er the river pois'd, the twain yet one, a moment's lull,

A motionless still balance in the air, then parting, talons loosing,
Upward again on slow-firm pinions slanting, their separate diverse flight,
She hers, he his, pursuing.

Whitman's wonderfully sensual, erotic imagery led to a widespread belief that bald eagles copulate in midair. In fact they don't but rather mate in a tree, on a rock ledge, or at the nest. And not all display flights are aquiline foreplay; many eagles in temperate climates perform nuptial displays virtually year-round.

Eagles devote considerable time to domestic affairs—nest building, incubating eggs, and taking care of flightless or otherwise dependent young. Much of their effort centers on the nest itself, either building a new one or, more commonly, tinkering with an old one. Both bald and golden eagles use the same nest or series of nests year after year, adding new tiers and weaving in new sticks until the eyrie reaches enormous size. One bald eagle nest in Ohio, in use for thirty-six years, measured eight feet across, twelve feet deep, and was estimated to weigh a ton. Another in Florida grew to almost ten feet wide and twenty feet deep.

Though they sometimes nest on mountain ledges or on the ground, both species prefer to build their eyries in trees. A conifer or tall hardwood with its top blown out seems especially attractive. Even new eyries are big, simply because eagles are big birds. A few springs ago a pair of bald eagles built one in a cottonwood along the Missouri River a few miles from where I lived; even as a first-year structure you could literally spot it a mile away.

Both sexes of bald eagles participate in nest building, and each pair usually will build two or three. The female golden does most of this work with only some help from the male; they build two to four nests, although at least one busy couple is known to have constructed as many as fourteen.

Egg-laying time is influenced by climate. It may come as early as March in Alaska or as late as October in Florida. Two eggs is an average clutch for both bald and golden eagles, with the second laid two to four days after the first. The females do most of the incubating, which takes from thirty-five to forty-five days.

The first egg laid hatches first, and the eaglet works many hours to chip its way free. At first it is weak and helpless, covered with fuzzy, grayish-white down. With extraordinary gentleness the female immediately begins offering tiny morsels of flesh with the tip of her beak. The young bird usually begins to feed vigorously on the second day and gains weight quickly. By the time the second egg hatches the first eaglet is strong and grown considerably larger. If there's only a two-day lag between hatches, both eaglets may survive. If it's four days, the youngest nestling is likely to meet a gruesome death.

The older, dominant one may attack and kill its sibling, may drive it to the edge of the nest where it dies of exposure to sun or wind, or may simply intimidate it to the point that it will accept no food and starves. A dead eaglet will either be eaten by the parent or fed to the older nestling. The pastoral scene of downy young snuggled companionably together doesn't happen in an eagle's nest.

The male adult does most of the hunting while the female does most of the actual feeding. For the first six or eight weeks she must tear up whatever prey her mate brings. The first of an eaglet's feathers begin to show when it's about four weeks old, and it will be fully fledged sixty-five to eighty days after hatching. It still cannot catch its own prey, however, and the parents continue providing food well after the young are able to fly.

Golden eagles look much the same at any age, though immature birds have white patches on the undersides of their wings and tails. Immature bald eagles do, too, including some white mottling on their legs, but their wing and tail patches are less distinct and are on the leading edges of the wings rather than down the center. Young bald eagles are often misidentified because they lack the white heads and tails of adults. Audubon in fact believed immature bald eagles, which he painted at least twice, were a separate species altogether.

The young grow a few white head- and tailfeathers after their first molt, and a few more appear each year. By the third year—or in some cases the fourth—they're dressed in full adult plumage.

Eagles' reproductive biology operates on a relatively long cycle, long enough that even the best-intentioned human interference can have a significant impact. The birds will not tolerate much human activity near their nests and do not renest if driven away from a clutch of eggs. Their mystique is such that management agencies have a hard time keeping well-meaning but disruptive birdwatchers away from nesting eagles.

Some other human attention, both direct and oblique, has proven even worse. Ranchers, sheepherders, and misguided "conservationists" have persecuted North American eagles for two hundred years, killing them under the banner of protecting livestock or game. In view of the facts, destroying golden eagles is no better than pointless; killing bald eagles for the purpose of protecting anything is sheer stupidity.

Although habitat improvements have led to increases in bald eagle populations throughout the Great Lakes region, there and in many other parts of the country eagles are still absorbing enormous amounts of organochlorine pesticides and mercury through the fish they eat. In Michigan, for example, where the number of nesting pairs has risen from 86 in 1980 to 268 in 1996, eagles fare better on the cleaner inland waters than they do on the more polluted waters of the Great Lakes. Among other effects, industrial and agricultural pollution reduces the birds' reproductive success, and populations in heavily contaminated areas are declining. Eagles have enjoyed considerable attention from conservation groups in recent years, although much of it unfortunately has focused on symptoms rather than on the real problems. Prosecuting those accused of killing eagles is necessary and admirable, but court proceedings don't bring back the dead and accomplish nothing at all to aid the many birds being slowly poisoned by pollution. Heroic rescue and rehabilitation efforts on behalf of individual eagles are admirable, too, and usually good for a fifteen-second sound bite on national news, but unless they result in a bird restored to the wild as a breeding animal they amount to nothing in terms of conserving the species and in fact represent a waste of resources that would be better used in behalf of habitat and environmental improvements.

Nature spends individual animals with a prodigal hand, but if, and only if, the survivors have a decent place to live and a supply of clean food and water, the species will prosper. Conservation efforts for eagles could be far more beneficial if their focus was broader. As goes environmental quality so goes the eagle—along with everything else, including ourselves.

Eagles are no more or less important in the fabric of nature than are the poorest worms that creep, but there's something about them that stirs the soul. My soul, anyway. It's been my good fortune to spend a lot of time in their presence. I lived for quite a few years near a federal wildlife refuge that's home to the largest population of wintering bald eagles outside Alaska and enjoyed special permission from the manager to hike into the part of the refuge where they live. I spent many a bone-chilling sundown sitting against a tree watching eagles tuck themselves in for the night right overhead, knowing I couldn't leave until full dark to keep from disturbing them, and happy to stumble quietly out on the edge of hypothermia just for the privilege.

It seemed, and still seems, a small price to pay. Especially when I now chance to see an eagle, disdainful of gravity, winging smoothly across the wind.

# 17

# Daylight's Dauphin

*I caught this morning morning's minion, king-*
*dom of daylight's dauphin, dapple-dawn-drawn Falcon in his riding*
*Of the rolling level underneath him steady air, and striding*
*High there, how he rung upon the rein of a wimpling wing*
*In his ecstasy! then off, off forth on swing,*
*As a skate's heel sweeps smooth on a bow-bend: the hurl and gliding*
*Rebuffed the big wind. My heart in hiding*
*Stirred for a bird,—the achieve of, the mastery of the thing!*

Gerard Manley Hopkins
"The Windhover"

My heart was not in hiding the summer it stirred for a bird. It was in Iowa. As usual in summer I was living with my grandparents on the farm. One of my jobs was to patrol Granddad's fields in search-and-destroy missions against burdock and buttonweed, and I was about it on a late-June afternoon, out at the far end of a hayfield east of the barn, rambling along with old Bonny, Granddad's dog, at my heels.

Shadows flickered past. I looked up to see a half-dozen pigeons winging over, part of a flock that regularly traded among the farms nearby. Granddad had issued standing permission that I was welcome to try my meager skills with a shotgun on them any time they came around, so I watched to see if they were bound for our place or the Herman farm farther west.

As they reached the center of the field something else caught my eye—a small, dark shape diving at them from above in a steep slant. Its speed was incredible, bulletlike, faster than anything I'd ever seen in the air. A moment later came a scream I can almost hear even now, and one of the pigeons seemed to explode in a burst of feathers. Instantly the shape sprouted pointy wings, executed a whiplash turn, and was back on the pigeon just as it tumbled into the timothy.

It was a literally breathtaking sight. I have no words to describe the feeling except to say in that moment I learned the meaning of awe in the truest sense.

Though I'd never seen one before, I knew exactly what that dark shape was; I'd pored over its picture in my bird books, read everything about it I could find, heard my father describe it as the fastest bird that flies. What a peregrine falcon was doing in southeastern Iowa I haven't a clue—unless the red gods simply decided to give a young lad something to remember for the rest of his life.

In those days the peregrine was commonly called duck hawk, just as the merlin was known as pigeon hawk, and as kestrels still go by the common name sparrow hawk. Hawklike they are; true hawks they're not. They and their relatives scattered around the world are falcons, and along with the caracaras they belong to a family all their own, the Falconidae.

They are an old family, having evolved to a distinctive form at least a million years ago, perhaps 5 million. About thirty-five species now exist worldwide; six are native to North America, and a couple of others show up as occasional visitors. The natives include the gyrfalcon, which lives all around the top of the world

and seldom ventures very far south; the peregrine and merlin, both of which are also found in Europe and Asia; the prairie falcon; the American kestrel; and the Mexican aplomado falcon. Eurasian kestrels have been recorded as accidental visitors in Massachusetts, New Jersey, British Columbia, and the Aleutian Islands. Northern hobbies have been sighted in the Pribilof Islands, the Aleutians, and at least once in British Columbia.

Falcons are not large birds. The gyrfalcon, largest of them all, isn't much bigger than a crow, and the smallest, the kestrel, is about the size of a blue jay. But even if they aren't as imposing as the big hawks and eagles, they are unquestionably the most magnificent of all the hunters of the air. Some are built for maneuverability; all are built for speed.

The typical falcon is the ultimate aerodynamic bird—a sleek, large head, heavy body, narrow tail, and long, slender wings pointed at the tips. These slim wings with their relatively small surface area require a rapid passage of air in order to maintain lift. Consequently the larger ones—like the gyr, prairie, and peregrine—simply cannot fly slowly. They can soar, but for active flight they have to maintain, as Hopkins put it, a hurl and a gliding and a big wind across their wings.

Even so, most falcons can hover, and kestrels are especially good at it. The Eurasian species, which is the falcon of Hopkins' poem, long ago earned the common name windhover, and it's not at all unusual to see American kestrels hovering over the grassy margins of roads and highways, looking for prey. The kestrels are also good at kiting—the trick of holding their wings steady to stay motionless against a strong headwind—which no doubt is the source of the kestrel's other old English common name, standgale. Our kestrel is the only North American falcon that regularly hunts this way, although the much larger rough-legged hawk is also good at hovering.

Few falcons are willing to turn down the easy opportunity for a meal of carrion now and again, but for the most part they're active hunters. What they're after varies somewhat according to species. Peregrines and aplomados prey almost exclusively on birds; merlins like birds, too, but also catch a great many insects. Not surprisingly for where they live, gyrs eat a lot of ptarmigan and ducks to supplement a main diet of small mammals. Prairie falcons prefer ground squirrels above all else, but they also kill ground-dwelling birds and a certain proportion of insects and lizards. The little kestrels pursue the most diverse diet of all—rodents, songbirds, insects, lizards, snakes, and just about anything else small enough to catch and dispatch.

Hunting methods vary as well. Most species are fond of working from a perch, taking flight when they've spotted likely prey. Some, like the merlin and gyr, attempt to close with prey from a high-speed cruising flight. Kestrels are essentially still-hunters, using a hover, kite, or perch as a vantage point. Prairie falcons often dive from great height to gain maximum velocity and then make a final approach on their ground-dwelling prey at low level. In active flight, the prairie is the swiftest of all the falcons.

The peregrine, however, is the most spectacular hunter, preferring to attack in a nearly vertical dive, called a stoop, its wings folded for least resistance from the air. Peregrines in stoop are said to reach a terminal velocity of almost two hundred miles per hour. Not having timed one with a radar gun I can't say it's true, but having seen it I don't find the numbers hard to believe.

In the actual contact the peregrine simply clobbers its prey with a closed fist, often including a passing rip from its single hind claw. The impact must be tremendous; it's enough to kill full-grown mallards, which are not only much larger but several times heavier than a peregrine. The pigeon I saw taken surely was dead before it fell two feet.

Any prey not killed from sheer shock is soon dispatched with a quick bite to the neck. One thing that distinguishes falcons from other raptors is a pair of toothlike scallops on their upper beaks with corresponding notches in the lower mandibles, like notches in the blades of poultry shears. They give the falcons maximum purchase for biting through neck bones.

As another bit of specialized equipment, the fastest-flying falcons have a series of baffles inside their nostrils. By impeding airflow these allow the bird to breathe while it's moving at high speed and at the same time protect its internal organs against the buildup of air pressure. Without them a prairie falcon or peregrine in full stoop would take in so much air that its lungs would burst.

While gyrs and prairies and peregrines rely almost solely upon speed, some of the others make use of agility as well. Merlins, for instance, are maneuverable enough to hunt swifts, swallows, and shorebirds by tail-chasing, and nimble enough to take dragonflies. Aplomado falcons catch a goodly number of bats.

Except for kestrels, which are distinctive in color and marking, falcons aren't always easy to identify from a distance because immature birds look different from adults, and some species display more than one color phase besides. At a glance, white-, gray-, and black-phase gyrfalcons look like three different species, though they're all *Falco rusticolus.* Peregrines and merlins also show three distinct forms. The peregrine races are called tundra, continental, and Peale's; the merlins are black, prairie, and taiga.

To complicate things a bit further, male and female kestrels and merlins show different markings. In the other species telling the sexes apart is very difficult—except that all female falcons are noticeably larger than the males.

This difference in size has evolved some linguistic differences as well. Among those who like to make fine distinctions, only females are called falcons; the males are called tiercels. Although this properly applies to every species, a few have their own terms. So for the gyrfalcon, the female is gyr, the male jerkin. For the merlin, it's merlin and jack.

All of this, and enough more to fill a hefty book, derives from the ancient sport of falconry. Hunting with captive falcons, eagles, and hawks has been practiced for at least 4,000 years, and over that time it has evolved its own specialized vocabulary and complex set of formalities—even to the extent that the falconer's choice of species once was determined by his social station. As *The Book of St. Albans,* published in 1486, puts it: "An Eagle for an Emperor, A Gerfalcon for a King, A Peregrine for an Earl, A Merlyon for a Lady, A Goshawk for a Yeoman, A Sparehawk for a Priest, A Muskyte for an holiwater clerke." (A "muskyte," by the way, was a male "sparehawk," or sparrow hawk, a European hawk different from a kestrel.)

In another carryover from the specialized lexicon, falcons are known by different names at different stages of their lives. Young birds are *eyasses* while still in the nest, *bowasses* from the time they leave the nest till they learn to fly, at which point they become *branchers.* Adults are *haggards*—which strikes me as especially apt for adults that have successfully reared a family of young. It's a busy, demanding time for them.

Probably because the female is larger, more powerful, and decidedly dominant, tiercels perform elaborate rituals of courtship with much bowing and scraping, gentle tempting and persuasion. As most species are solitary and antagonistic during the rest of the year, the courtship serves to establish what must necessarily be a cooperative, if brief, relationship.

Preferred nesting sites vary somewhat from species to species. Peregrines, gyrs, and prairies are mainly cliff nesters. Aplomados often take over abandoned tree nests built by other birds. Merlins like tree cavities but also frequently nest on the ground. Kestrels use the greatest variety of sites—tree cavities, niches in cliffs, crevices in buildings, even nest boxes where they're available.

Regardless of where it's located it won't be much of a nest—most likely none at all, for falcons are content to lay their eggs on whatever surface is available. The number of eggs varies a bit, too, but three or four is a general average. Like other raptors, falcons do not rear big families.

◆

Females do most of the incubation, while the tiercel's job is to fetch food. He will continue to do this for some time even after the eggs hatch, feeding both the female and the young. Sometimes he brings food directly to the eyrie, or nest—which might earn him some lumps if she's feeling out of sorts. Frequently though, she flies out to meet him, and he hands over the goods in midair. These exchanges involve some impressive aerobatics.

About twenty years ago when I lived in a village in northwest Missouri, a pair of kestrels nested in the attic of the house across the street, giving us several weeks of splendid entertainment. I don't know if all kestrels exchange food the way these two did, but in their favorite method the tiercel would hover high over the open lawn with his legs down, talons holding whatever he'd caught. She'd fly up, make a couple of loops to get the range and then swoop in and take it, sometimes with her beak, sometimes with her talons, turning over in flight. They never seemed to fumble. In fact the only mishap I saw came one day when the little chap showed up dangling a two-foot garter snake. She caught it okay, but he was a mite tardy letting go and got a good jerk in a momentary game of crack-the-whip.

Except for the kestrel, which is both plentiful and widely spread, the chance to see any of the falcons is at best a sometime thing. Gyrs rarely visit the United States, and aplomados are only casual visitors in a narrow strip along the Mexican border. Peregrines and prairies and merlins aren't really abundant anywhere, even at the best of times—and the latter twentieth century has not been the best of times for falcons.

Loss of nesting habitat has been severe over most of the country. The most insidious threat, however, began to appear right after the Second World War with the development and use of organochlorine chemicals as pesticides—DDT, BHC, aldrin, dieldrin, and heptachlor. Because these poisons are biocumulative—that is, they accumulate in animal tissue, especially fat—they gradually started traveling up the food chain, building to ever-higher levels as they neared the top. As falcons certainly are the highest link among predatory birds, they suffered just as much as the songbirds that live on a diet of insects. It took longer for the effects to show up, but by the 1960s the evidence was clear—dead adults whose bodies were loaded with poison, reproductive rates dwindling because of inhibited sex hormones, nesting failures as a result of thinning eggshells, on and on.

The peregrine was hardest hit. Populations steadily declined worldwide, disappearing entirely from one country after another. In 1972 the United States banned DDT entirely and began phasing out the use of dieldrin and aldrin, but by the late '70s the North American population still showed no signs of recovery. A number of institutions, notably Cornell University and the Canadian Wildlife Service, launched projects for captive breeding and release.

In the years since, these programs have achieved good levels of success, and efforts continue even now. Whether they truly have begun a renewal of peregrine populations or simply are a holding action against the inevitable, only time will tell.

Falcons are no more nor any less valuable to the natural scheme than meadowlarks or mice. It's one thing if nature in its own way decides that falcons, or any other animal, should cease to exist; extinction is an inevitable component in the evolution of life. But to allow human cupidity or willful ignorance to wipe out any species leaves us much to answer for.

Some things are of ineffable value simply for the way they touch our spirits. To lose them is to lose a part of ourselves that we can scarce afford to be without. I'd hate to see the day when no young person's heart could ever again stir for a bird or feel the sheer wild thrill of witnessing the achieve of, the mastery of the thing.

◆

# 18

## Night Hunters

The birds felt they needed a king and agreed that whoever among them could fly highest would thereafter rule them all. The eagle outreached the rest, but as he hung there against the sky, his strength finally spent, the wren, who had ridden unseen upon the eagle's tail, rose up and climbed higher still.

Angry at such deceit, the others confined the wren to a mousehole and set the owl to stand guard until a proper punishment could be decided. As the debate dragged on, the owl grew drowsy, fell asleep, and the wren escaped with his misdeed intact. Burdened with shame, the derelict sentinel has hidden himself from view ever since, and that explains why the owl is a creature of the night.

Or so an ancient central-European folktale has it. An equally old story from Brittany says the owl was sentenced to a solitary, nocturnal life for refusing to join the other birds in contributing a single feather each to the wren, who somehow had lost his own.

Unflattering tales, but predictable enough, for man, the perennial heliotrope, has always mistrusted the dark. Even now, at some atavistic level, we tend to doubt the virtue of any creature that shuns the light of day.

The owl, of course, didn't choose its nighttime life but rather evolved to fill a niche that offered a prosperous living to any creature able to operate efficiently after sundown. And it did so a very long time ago, some 60 million years before the present, at about the same time the dinosaurs were making their final exit. *Protostrix mimica,* known from fossilized remains found near Worland, Wyoming, is the earliest owl so far discovered. When it was alive, *Protostrix* shared the world with the last of the great reptiles and with the first flourish of mammals—one of which, a lemurlike creature with grasping toes, would in the far-distant future become man.

The descendants of *Protostrix* evolved into modern forms, which probably originated in Eurasia, and into a surprisingly large, relatively varied family that thrives on every continent except Antarctica and on a great many oceanic islands as well. In historic times, 137 species of owls have been identified; four of these, along with a handful of subspecies, have gone extinct since about 1730.

Taxonomically, owls make up the order Strigiformes and are grouped into two main families—Tytonidae, the 10 species of barn and grass owls; and Strigidae, the huge family of "typical" owls that comprises the other 123 living species.

As you might imagine, animals that have developed into so many different forms also come in a great many sizes. The pygmy owls of the genus *Glaucidium* are the smallest, averaging about six inches from tail to top; the smallest of all, the least pygmy owl of Mexico and the Amazon Basin, scarcely measures five. At the other end of the scale, the largest of the *Bubo* species, the eagle owls, commonly reach thirty inches or more. Owls of all sizes are distributed more or less evenly around the world, and with one exception both the smallest and the largest are represented on every continent where owls live. Only Australia has neither a *Bubo* species nor a pygmy owl—although at twenty inches or so, the Australian hawk owl is about the size of the smaller eagle owls.

Such differences, however, are only a matter of degree, for no matter whether it's as short as a lightbulb or as tall as a tire, an owl is unmistakably an owl. Their heads are large, their tails short, their eyes big and forward-looking, set close together among feathers that form a broad facial disc. Their beaks, not very large to begin with, curve downward and appear even smaller because they're partially hidden by the facial plumage.

Some hawks' and eagles' perching posture is about equally upright, but you'll know an owl even from a distance if a second look shows a rounded, almost body-wide head that appears set directly upon the bird's shoulders without benefit of neck. Distinctive birds, owls are, like penguins.

Something in the almost-humanlike face and double-barreled gaze of an owl tells you this is an animal of serious intent. Or as William Service puts it in his delightful little book titled *Owl*: "The animal which looks back at you with two eyes at once tends to stand high in the local food chain."

And so it is with owls, for every one of them, large and small, is a hunter. The prey might be no bigger than a moth or it might be a full-grown cottontail rabbit. It might be an earthworm, a frog, a bird, a fish, brown rat, lemming, or field mouse. Big owls aren't even averse to eating little ones. Such cannibalism is probably rare, however, for even though big and little owls frequently share the same habitats, their differences in size—implying by extension some differences in the size of the prey they seek—tend to help them stay out of one another's way.

Size differences also allow them to exploit a broad spectrum of available food, but timing is the most important key to the owls' success as a race of predators, for most make their living while the competition is asleep.

Which isn't to say that all owls are nocturnal. The seven species of African and Asian fishing owls work almost exclusively by day, as do the dozen or more species that live in the far north, where the summer sun either never sets at all or offers nothing more than a few minutes of twilight each day. Burrowing owls, pygmy owls, some eagle owls, and several others are either decidedly diurnal or at least crepuscular. Altogether, about two out of every five species worldwide live more or less diurnal lives, and almost any owl will hunt by day if it's hungry enough.

Still, the classic owl is a nighttime hunter, and as such they have managed to fill that niche better than any other bird.

The nocturnal lifestyle confers two important advantages. For one, it allows owls to go about their business without competition from the raptors that hunt by day—which works out well for both. In fact, certain North American owls and hawks literally replace each other at daylight and dark, working over the same food supply and the same habitats all around the clock. This avian shift-work dovetails especially neatly for the screech owl and sparrow hawk, great horned owl and red-tailed hawk, short-eared owl and marsh hawk, and the barred owl and red-shouldered hawk.

Not that owls have the nighttime sky all to themselves. Nightjars and bats also are nocturnal hunters, but they're insect-eaters and therefore offer no serious interference. Although owls as a class take a remarkable variety of prey animals, more owls eat more rodents than anything else, and that, too, makes night-work especially profitable.

Comprising about 3,000 species worldwide, rodents are the most diverse and numerous of all the mammals. With few exceptions they're of a size to be food for nearly any owl, and most of them are chiefly active at night. Like the Renaissance naturalists who referred to any animal smaller than a rabbit as a "mows," owls take a supremely democratic attitude toward the teeming world of small, furry creatures. Where a taxonomist might see lemmings, voles, hamsters, gerbils, shrews, mice, or rats in a multitude of forms, an owl simply sees dinner.

The various predators whose business lies in harvesting this bounty have all developed skills and tools specialized to the task, but none is more splendidly equipped than the night-hunting owls.

Contrary to the more persistent folktales, owls cannot see in absolute darkness nor are they blind by day. The facts are considerably more impressive. Owls' eyes are different from those of hawks and eagles. The corneas are larger, the lenses wider and thicker, and the optic chambers somewhat flattened to bring the retinas close to the lenses. The retinas are richly supplied with light-sensitive rod cells, and the whole system combines to create a window large enough to gather every available photon and to resolve images over a short focal distance, thereby preserving maximum brightness.

Although visual acuity and photic sensitivity tend to be mutually exclusive, the owl's wonderful ability to see in very dim light apparently has not evolved at any great expense to the detail of what they see. There is, in fact, evidence to suggest that an owl sees better by night than we do by day.

Neither the owl nor any other animal can see in complete darkness, but complete darkness occurs only deep underground and in manmade structures, never in the open-air world. A moonless, heavily overcast night might seem completely dark—and where our comparatively poor eyes are concerned, it might as well be—but the darkest night probably is no dimmer to an owl than a cloudy day is to us, for an owl's ability to use available light is about a hundred times better than ours.

Owls are well equipped for daytime vision, too. Their irises can stop down almost to pinpoints, and in extremely bright light owls can cover their eyes with translucent nictitating membranes that shield against the glare while still preserving the ability to perceive movement and shape.

◆

Because owls' eyes are frontally placed, their field of vision isn't particularly wide—only about 110 degrees compared with 340 degrees for a pigeon and 180 degrees for a human. Nonetheless, owls have an exceptionally wide field of binocular vision, which is the portion of the total field of view covered by both eyes at once. A pigeon's binocular field is only 20 degrees, ours is 140, and an owl's is 70, which is far wider than that of any other bird.

This obviously is a great advantage, because binocular vision is three-dimensional, allowing depth perception and accurate judgment of angles. For the owl, however, it is not a wholly perfect system. Most birds can move their eyes to some extent, but owls cannot. Owl eyes are irregularly shaped, for one thing, and are extremely large, for another. Movable eyes require a proportionately large skull and a complex system of muscles—no handicap to a land-bound animal but an unnecessary burden in both size and weight for an animal meant to fly. Since evolution favors characteristics most useful in survival, owl eyes have developed to maximum reasonable size and sensitivity but, as a trade off, are firmly fixed in their sockets.

An owl therefore cannot adjust the convergence of its eyes at every point between close-up and infinity, as we can. Instead, owls' vision converges at only one point, so any object closer or farther than that no doubt registers as a double image. In compensation, nature has equipped owls with extremely flexible necks, and like some other animals, they use motion parallax to help them judge distance and angle. An owl interested in some object bobs its head up and down, side to side, even turns its head almost upside down, all to discern a precise location out of the relative movements among the things it sees. To us, it looks like a charming display of curiosity; to an owl, it's a matter of getting several points of view on something that might either be dangerous or good to eat.

That remarkably lithe neck also permits the bird to rotate its head through about 270 degrees of arc, which is where we get the bit of folklore that says you can kill an owl that's watching you by walking in circles around it until it wrings its own neck.

Owls' sense of hearing is as highly developed as their sight, especially in those northern species that are most persistently nocturnal. Fifty-seven species wear ear-tufts of feathers; these may be long or short, stand straight upright or slant outward from the head, but in no form do they have the slightest thing to do with hearing. An owl's ears are half-moon-shaped openings on each side of its head. They are enormous, in some species nearly girdling the skull. Besides the facial disc of feathers, which may act as a sort of parabolic reflector for gathering sound waves, the ear openings are surrounded front and back by mobile, feather-covered flaps of skin, which the birds can adjust in a variety of ways to pinpoint the location of potential danger or prey.

Their hearing is particularly well-tuned for high-frequency sounds—the high-pitched squeaks of rodent voices and the faint patter of feet as they hustle through grass or fallen leaves. Researchers have discovered that barn owls, the finest of all the nocturnal hunters, can successfully catch prey in absolute darkness if they're given some auditory clues. Most owls, in fact, are keen-eared enough to home in on prey within one degree of accuracy, navigating on hearing alone.

They're able to do this because their ear openings and the surrounding flaps are asymmetrical. One ear, usually the right, is larger than the other, perhaps as much as half again, so owls perceive certain sounds better in some directions than in others. At the all-important high-frequency ranges, their hearing is most sensitive when their eyes are pointed toward the sound source. The higher pitched the sound, the narrower this region of maximum sensitivity.

The system, moreover, works in all three dimensions. The asymmetry of their ears not only ensures that the hearing in one is a mirror image of the other but also allows owls to track objects moving straight toward them or straight away, because one ear is located slightly higher on the head than the other. So by orienting its head to equalize the sound reaching its ears, an owl can maintain a very precise course toward what it hears.

It might do so in a long swooping glide or with continuous wingbeats, but in either case the prey won't hear a thing until it's too late, for the flight feathers that most species wear are designed to make no sound as they cut through the air. In most birds, the primary and secondary wingfeathers are hard-edged; the

system of interlocking barbules and barbicels extends fully to the tip of each barb. The barbs of owl feathers interlock only partway out, leaving the edges softly fringed and silent. Making its living as it does, soundless flight is as useful to an owl as good vision and hearing, since rodents have sharp ears, too.

◆

To a mouse or a vole, the business end of an owl must be a dreadfully fearsome prospect—huge powerful feet wearing big, curved talons sharp-tipped as locust thorns. Of the four toes, two point forward and one behind, but the other is on the outer side of the foot, oriented sideways. The bird can use it either as a hind toe for perching or as a side toe to widen its grasp for snatching or holding prey.

Once they have a morsel in hand, owls aren't much given to leisurely or dainty dining. Straight down the hatch it goes in a mighty gulp. While an owl might decapitate and strip a few feathers from the birds it catches, rodents go down fully dressed; everything is usually swallowed whole, fed into a remarkable digestive system that has neither crop nor gizzard. A few hours later, after its stomach juices have extracted every smidgen of soft matter, the bird blinks once or twice, gapes its beak, and regurgitates the leftovers in one tidy package.

These castings, often called owl pellets, contain everything the bird can't digest—bones, teeth, beaks, toenails, insect carapaces, fish scales, even the tiny, chitinous bristles from earthworm bodies—all neatly wrapped in feathers or fur. They're perfectly clean, dry quickly, and amount to a *post facto* menu of an owl's last meal. Some species can be identified by the size, shape, and color of their castings.

Owls usually produce castings during the day while they're roosting, and since some species tend to use the same roosts for long periods of time, pellets can accumulate in impressive numbers. Dismantling one is more fun than rummaging for the prize in a Cracker Jack box. Soak it in warm water for a few minutes, lay it on some blotting paper, carefully pick it apart with tweezers and a dissecting needle, and you'll soon have a nifty collection of little bones, skulls, and other detritus that speaks eloquently of nature's lovely cycle of life and death on a miniature scale.

Besides providing a pastime for the insatiably curious, owl pellets are evidence of an extraordinarily efficient organism, and this is good, because for all their splendid predatory skills, owls don't catch everything they attempt. Even if they did, they'd still harvest only a fraction of the prey-animal population. No animal can eat what isn't available, and nowhere is that more profoundly clear than in the relationship between predator and prey.

Populations of both ebb and swell in a cycle of harmony and counterpoint. When circumstances combine to produce abundant prey, predators flourish. An abundance of predators in turn contributes to—but never solely causes—an eventual population crash among the prey, and predator populations quickly decline in response. Continual, cyclical change is the nature of nature.

Owl populations rise and fall like all the rest. When rodents and other prey animals reproduce well, owls do, too. In leaner years, they lay fewer eggs and sometimes, when prey shortage reaches a truly critical point, fail to nest at all.

Springtime is the usual breeding season, but some species, notably barn and short-eared owls, occasionally nest during mild winters when hunting is especially good. Whenever it happens, the timing is such that the young will hatch just at the start of the annual period when the maximum amount of food is available.

Even under the best conditions, though, mating is a harrowing affair. Through most of the year owls are solitary, territorial birds, not particularly fond of company and even less tolerant of competition in their hunting grounds. Consequently, a male owl in a conjugal frame of mind faces a few problems: His lady friends are bigger than he is (true of almost every owl species), every bit as well-armed, and may be in a humor to shoot first and sort things out later. Except for snowy owls, both sexes look alike, so the poor chap can't even rely on his appearance to let her know that he'd rather be a suitor than a casualty—nor for that matter can he rely on hers as a means of knowing whether she's a potential mate or another male who'll be decidedly unenthusiastic about an amorous come-on.

So owls use various behavioral means of advertising their identities and intentions. They establish sex recognition by calling back and forth to one another, and then the males perform courtship displays that involve much beak-snapping, feather-ruffling, swaying, bowing, and other dancelike maneuvers. Like other animals, from Lycosid spiders to finches and hawks, male owls often sweeten their propositions with gifts of food, which serves both as additional nutrition to help the female produce high-quality eggs and also to distract any thoughts she might have about eating him.

The nest may be a tree cavity, a ledge or crevice in a rocky cliff, a barnloft, the attic of an abandoned farmhouse, an old hawk or crow nest, or an underground den dug by a prairie dog or some other animal. For a pair of snowy owls it might be nothing more than a depression on the open tundra. In any case, the birds won't do much home improvement work.

The number of eggs she lays—perhaps only one or as many as fourteen—depends upon her species and the food supply. The eggs are white and roundish, and in most species the female begins incubating as soon as the first one is in the nest. The embryos take three to five weeks to develop, and hatching is typically asynchronous, which is to say the eggs hatch over a period of days rather than all at once.

If food is abundant and no accidents befall the adults, the whole brood stands a good chance of surviving, even the latecomers. Otherwise, the youngest and weakest starve while the older ones prosper. Those that do survive grow quickly, soon develop flight feathers, and leave the nest to get on with the business of learning to be owls.

Through it all, the adults are conscientious parents, kept busy with the demands of feeding themselves and a hungry brood, all the while keeping a weather eye for anything that might be a threat to the young. The ground-nesting species use distraction displays to lure intruders away, and many others put on fearsome acts of intimidation. These are no idle threats, because any owl defending a nest is one tough customer. It will attack almost anything in a flurry of thrashing wings and wickedly sharp talons and will press the fight as long as it takes to make the point. Humans are by no means immune to a drubbing or the odd gash now and again, so if you know the location of an owl nest, it's wise to give it a wide berth. If you're suddenly swooped by an owl for no apparent reason, you can bet there's a nest nearby.

That silent, unexpected rush out of the night can startle you half out of your wits, and this as much as anything probably has contributed to the scapegrace reputation owls have among some people. Had it happened to me before I had some good experience with owls, I might not be so fond of them myself. But I am, and it goes back a long way, to my childhood and Mr. Ayres.

He was a local attorney and a friend of my father's and an amateur ornithologist of great repute. Between the male cardinal that lived for years in a big flight cage at one end of the dining room, the dozens of photographs and study skins around the house, and his delight in sharing what he knew about birds, an hour with Mr. Ayres was better than a day at the circus.

One late-winter Saturday, Dad suggested that such a visit would hold a particular treat. He was quite right, because Mr. Ayres showed me the most impressive animal I'd ever seen.

It was pure, pale white flecked with black and stood two feet tall in a small wire cage, talons the size of my fingers circling a wooden perch as thick as my wrist. Mr. Ayres explained that it was a snowy owl, a bird of the Arctic tundra, and that occasionally, driven by hunger when the lemming population dies off, they migrate south, sometimes reaching as far as Missouri and the Carolinas. This one, no doubt hunting rabbits in the southern-Iowa countryside, somehow got tangled in a wire fence and stayed trapped for some time before the landowner found it. There were no raptor rehab centers in those days, and after the local game warden extricated the bird he brought it to Mr. Ayres for convalescence.

So there we stood, face to face, this ineffably beautiful creature with great yellow eyes and walrus mustache, and a speechless, equally wide-eyed boy of ten. It gave me precisely the same feeling I got years later, standing on a windy, rock-tumbled headland on the coast of Oregon and watching a pod of gray whales rise and blow just a few yards away. It was, to use the true sense of a sadly misused word, awesome.

The owl was completely calm and looked at me with a gaze that went somewhere near the bottom of my soul. Mr. Ayres opened the cage door; I reached in and gently stroked the back of its head as I might have stroked my dog's. The bird slowly closed its eyes, as if pleased to the point of dozing.

I didn't say much on the way home, and Dad, who always seemed to know what I was thinking, simply said, "I expect you're the only guy in town who petted a snowy owl today."

I expect I was, and my memory still holds it as a high point in a lifetime spent fooling with animals. So, too, are the foundling screech owls that ruled several weeks of my life more than twenty-five years later.

I was working for a fish and game agency at the time, and the state ornithologist came by my office one April afternoon with a boxful of owls. The city grounds crew, taking down a dead tree in a local park, had discovered the nest, so as is the lot of ornithologists everywhere, my friend Jim found himself with eight hungry owlets on his hands. They were old enough to have a good chance at survival but too young to fend for themselves, and Jim parceled them out among staff members willing and able to have a go at rearing them for release.

Two took up residence in our spare bedroom that night, two baseball-sized mops of white down fitted with big eyes and enormous feet huddled together in the little nest box I built. And thus the fun began.

Federal and state law provides that only certain qualified people may possess wild raptors for any reason, and this is good law, because the needs of raptors are extremely difficult to fulfill.

Just keeping them properly fed is demanding enough; preparing them to survive once released makes the task harder still.

Parent owls tear up the food they give their nestlings, so for the first couple of days mine ate bits of stew meat, gristly as I could get, from my fingers. To provide vital roughage I wrapped these morsels in dog hair, which is never in short supply at my house. But such cobbled-up stuff was only a temporary measure, because to an owl roughage means more than having bran flakes for breakfast. It means fur, feather, and bone, and for me it meant an all-out campaign. I spent an hour or so every day hunting my neighbor's barnlot, potting sparrows and starlings with a .22 rifle, chopping the carcasses into increasingly larger pieces as the owlets grew. I trapped mice by the bucketful in a friend's pigeoncote and once when the wild supply failed, even bought out a petshop's inventory. Owls with a whole nestful of voracious appetites clearly do not have time for hobbies or weekend vacations.

Tail- and wingfeathers began showing through their down within two weeks, the rusty color indicating that both would be red-phase birds (gray is the other typical screech owl color), and by then we'd all grown fond of one another. Imprinting is common among animals, and it's one reason why human-reared foundlings are seldom successful in the wild. Deprived of contact with their own kind while young, they fail to develop an instinct for avoiding people, which almost invariably causes them grief later. Consequently I tried to keep my own contact with the owlets to a minimum, hoping they would identify more with each other than with me. It wasn't easy, because screech owls are amiable little chaps.

From the beginning, they were willing to perch on my fingers, although if I startled them by suddenly looming up in front of their box, they'd hiss and snap their beaks in a typical owl gesture that sounds as much like a typewriter as a warning. (Lord knows what they thought I was doing in the next room most evenings.) William Service describes the screech owl he reared as an animal the size of a beer can with the personality of a bank president, and that sums it up nicely. Mine were fond of nibbling my beard and my ears and enjoyed having their heads and necks tickled, but they didn't want to be grasped or picked up bodily. An attempt at that would provoke a struggle, complete with beak-snapping and perhaps a squeeze from needle-like talons, just hard enough to let me know that owlish dignity has its limits even for two ragamuffin specimens wearing a combination of rapidly growing adult plumage and remnant tatters of nestling down.

Because they were virtually identical in size when I got them I knew they were close in age as well, no more than two days apart, which was a guess soon confirmed. One always was two days ahead of the other, both in growing feathers and in developing skills. Flight, for instance. Ill-prepared to use their parents' method of teaching flight by example, I tried another approach that worked surprisingly well: With a bird sitting on my finger, I simply lowered my hand a foot or so, just quickly enough to make it think it was falling. After a couple of sessions and some tentative wing-flapping, the older one simply lifted off my hand, flew across the room and clung to the top of a window frame looking surprised. But then owls always look surprised. Two days later the other did the same, and after that I gave them free run of the room for an hour or two every day.

As a sort of halfway house toward independence, I built a big flight cage on the deck, with tree branches for perching, sheltered places to doze through the day, and a wide floor covered with straw, into which I released several mice each day to give the owls as much practice as possible at catching their own food.

Finally, just at twilight on a quiet June evening, I opened the flight-cage door one last time. As usual for that time of day they were beginning to stir, and one, the older, presently flew over and perched on the sill. After a minute or two of sizing things up in its head-bobbing way it flapped silently out and arced over the roof toward a big oak at the far corner of the house. The younger one sat on the deck railing for a while and then headed east across the pasture. I saw it settle into a treetop, a tiny silhouette against the sky. I watched until full dark, wishing them well and feeling a mixture of loss and hope.

I never knew whether they survived. Even after too much time had passed for them to be still alive, I thought of them every time I heard a screech owl call in the timber nearby and always wondered if the bird I was hearing had an ancestor who learned to fly in my house, from my hand.

◆

# 19

# Singer in Scarlet

At the house on the edge of town, where I grew up, the back yard ended in a patch of woods, a few acres of southeastern-Iowa countryside that seemed to me a vast and enchanting wilderness. It was a youngster of a place, as woods go, regrowth of thirty or forty years' duration, some parts of it dense and brushy, others sun-washed, grassy, and dotted with ranks of sumac—a variety of habitats for a variety of small animals. It was home to a fair complement of rabbits and squirrels, a few possums, now and then a transient raccoon, but mostly it was alive with birds. I spent many hours poking around those woods, and by the time I was eight or nine, I'd made some headway at matching what I saw with the pictures in the bird book my parents gave me.

Two in particular intrigued me no end. I was always on the lookout for the little screech owls that liked to fly out of the woods at night, sit in the elm tree outside my window and sing their eerie, quavering songs like souls lost in the dark. They were easy enough to see then, peering round-eyed and astonished-looking into my flashlight beam, almost impossible to find next day, no matter how many big old trees I studied limb by limb.

The other was a daytime bird, as easy to spot as the screech owls were difficult, shining like scarlet embers against the lush green of spring and summer, like a crimson spark in the snow. Even now, I cannot let a cardinal pass without a second look.

Nor could my father and grandfather, who seemed to comment upon—and point out to me—every "redbird" they saw. To them, a cardinal was something special. At the time, I thought their interest, like mine, was simply that brilliant flash of color, eye-catching even when the birds are abundant. I didn't know until much later that both of them were old enough to have known a time when seeing a cardinal near home was a rare thing indeed.

As it happened, the cardinal had come to Iowa only a relatively short time before I did. Until the latter nineteenth century, it was strictly a southern bird, rarely seen north of the Ohio River valley; the 1886 National Museum checklist noted it as only a casual visitor beyond the northern boundary of Kentucky. But then, for a variety of reasons, *Cardinalis cardinalis* developed a pioneering spirit.

By 1895 its range extended to the Great Lakes. The first recorded Canadian sighting came the following year, on November 30, 1896, at London, Ontario, and by 1914 the cardinal was a permanent resident across southern Ontario and into the southern part of the Hudson River valley. Although the harsh barrier of winter climate effectively halted their advance at about the forty-eighth parallel, the birds eventually worked their way eastward to the Atlantic coast and as far north as the southern tips of New Brunswick and Nova Scotia.

The colonizing proved easier in the Midwest. From about 1890 through the early 1930s, cardinals pushed steadily and quickly up the Mississippi and Missouri river valleys, moving into Illinois, Iowa, Wisconsin, Minnesota, and westward across the Great Plains. Sizable populations naturally took a while to develop, even in the southern portions of the new range. In Iowa, according to the literature, eight observers counted 36 cardinals during a Christmas census in 1923; at Christmastime, 1929, seventeen observers reported 149. By 1945, when I was born, they were plentiful in our corner of the state, but to my grandfather, born in Illinois in 1880, and to Dad, born in Iowa in 1908, the redbird was always new.

What prompted the cardinal to spread out from its ancestral southern range no doubt was the same thing that induces any animal to colonize new territory or to abandon its old haunts—habitat. In the 1880s and '90s timber cutting leveled much of the great northern forest, creating in its wake new habitats as the woods began to regenerate themselves in brush and saplings. At about the same time mechanization spread all across farm country. Grainfields grew larger. Old-growth woodland diminished, and some of it regrew as brushy field edge. Everywhere, the land supported vast new acreage of seed-bearing plants, both cash crops and annual weeds. All of it, from New England to the Plains, became precisely the sort of habitat where the cardinal can flourish.

Throughout its range, which now extends all the way to Central America, the cardinal is a bird of the woodland edge, favoring a mixed landscape of open timber, young trees, brush, dense shrubbery, evergreen thickets, canebrakes, viny tangles of honeysuckle, wild grape, trumpet vine, Virginia creeper, or almost anything else that offers food and shelter.

A cardinal's physical equipment implies something of both its lifestyle and its family tree. Its feet, slender, with three toes before and one behind, are those of a percher, its heavy, cone-shaped bill that of a seed-eater. The cardinal shares these characteristics with its relatives in the family Emberizidae—the grosbeaks, buntings, and sparrows—and with its cousins, the finches. Like many of its kin, the males are brightly clad, the females and the juveniles of both sexes more drably dressed.

Ornithologists have long been as fond of cardinals as the rest of us and until fairly recently insisted upon maintaining distinctions among a half-dozen subspecies: the Florida cardinal, native to Florida and southeastern Georgia; the Louisiana cardinal of southern Louisiana and southeastern Texas; the gray-tailed cardinal of Texas, Oklahoma, and Mexico; the Arizona cardinal found in the southern parts of California, Arizona, and New Mexico; and the Santa Gertrudis and San Lucas cardinals of Baja California. Since only slight differences in size and color distinguish these from the more common northern specimen, they're now considered simply local representatives of a single species.

Taxonomic nitpicking aside, there's no mistaking a cardinal, particularly a male with his black mask and bib and brilliant crimson livery. You wouldn't confuse him even with his closest relative of all, the pyrrhuloxia, *Cardinalis sinuatus,* common throughout Mexico and in parts of the Southwest. Although commonly called the gray cardinal, the pyrrhuloxia has a noticeably larger, almost parrot-shaped beak, and not even the males are so thoroughly, earnestly red.

Part of the cardinal's stage presence stems from the fact that he is genuinely red, for red is one of only three true pigments found in feathers (yellow and brown are the others). All other bird colors, including iridescence, are a function of reflected light. A bluebird, for instance, is blue only when the light bounces off his plumage, but a cardinal is red all the time.

He behaves like a red bird, besides: a tough guy with a slightly frowny look and cocky attitude. The frown is an illusion, created by the black patches around his eyes; the attitude isn't.

Watch the cardinals at your feeder this winter, and you'll notice that while the females are mild-mannered, most of the males are belligerent little thugs. They're sociable to the extent of forming into flocks during the winter, groups of a half-dozen to twenty birds that include both sexes in about equal numbers. They tend to move around together and to feed at the same times, early morning and late afternoon, but once in a weedpatch or on a feeding tray, the males get as peckish as a pack of jackals. They won't tolerate females feeding next to them, not even their own mates, and there are constant little squabbles as they displace one another from choice spots. Occasionally, if three or four go for the same spot at the same time, there's a momentary free-for-all that looks like a scarlet pinwheel.

Come the middle of March, though, the little ruffians clean up their act considerably, at least where the ladies are concerned. Not only do they allow the females space at the feeder but often actually hand over choice tidbits, beak to beak. Within a few weeks, they'll be singing together, and the males will perform a little courtship dance, landing on her branch and hopping sideways in her direction, their necks stretched out and their crests fully raised.

For all their attentiveness and good behavior as suitors, however, the males are even shorter-fused among themselves during the breeding season than they are through the rest of the year, frequently scuffling over females or, once mated, over territory.

If no one else is handy, they often pick fights with their own reflections in windows, glass doors, or shiny hubcaps; the actors in these little comedies usually are males, but females do it, too, now and then.

More often, though, the females are busy with nest building. She might choose any kind of shrub, bush, vine-tangle, or small tree, but she'll almost always pick something dense. Where I live, cedars are quite popular, with blackberry thickets, hawthorns, and locust trees distant runners-up.

Wherever they decide to build, the nests generally will be within six or eight feet off the ground, and they'll hold eggs by early April. The clutch might number only two or as many as five; three or four is most common. The female handles virtually all of the incubation chores, which is why she isn't nearly so brightly colored as her spouse. He meanwhile busily searches out food and either brings it to her on the nest or at times appears to lead her to whatever source of food he's found.

The eggs hatch twelve or thirteen days after incubation begins, and then both adults industriously feed the young. Although about 70 percent of the adults' diet comprises seeds and fruits and other vegetable matter, about 90 percent of what they give their young are insects—cicadas, grasshoppers, beetles, caterpillars, crickets, grubs, fly larvae, and just about anything else they can catch, including a good many insects that attack domestic and agricultural crops. The feeding instinct is so strong that male cardinals whose nests were destroyed have been known to transfer their feeding efforts to robin hatchlings in nearby nests.

The nestlings prosper on such protein-rich food and leave the nest when they're nine or ten days old. They're capable of some flight at that age, though it may be little more than fluttering. Those that do survive will be flying strongly in another week or so.

The parents continue feeding the young for a couple of weeks after they bail out of the nest, but as with most passerine birds, childhood among cardinals is short. By about six weeks after they hatch, the young ones are completely on their own, and the adults are starting over with a new nest. A mated pair normally bring off three broods, sometimes four, in a single season.

Such enthusiasm for family rearing would soon have us neck-deep in cardinals were it not for a fairly high mortality rate among the young, which is true of nearly every species of small bird. Sometimes trouble comes early. Where their ranges overlap, cowbirds are especially fond of stashing their contraband eggs in cardinal nests. Cowbird eggs hatch sooner than most other birds'; the unwitting adoptive parents feed the hatchling intruder, and by the time their own eggs pip, the cowbird chick is big and strong enough either to kill its nestmates or simply starve them out by monopolizing the food.

Black rat snakes, splendid tree-climbers and egg-eaters that they are, have little trouble plundering cardinal nests. (But don't rush off for a hoe or a stick if you see one; blacksnakes are not important songbird predators, and they are superlative mousers, better than a whole battalion of cats. A handful of eggs is scant tribute to pay for having one around.)

House wrens sometimes puncture cardinal eggs, and blue jays occasionally wreck both eggs and hatchlings—all presumably motivated by an instinct to thin out the competition for food. Shrews, squirrels, housecats, snakes, weather, automobiles, and plate-glass windows also exact a toll on songbirds, cardinals included, at every age from egg to adult.

Mites, ticks, lice, and other typical avian vermin parasitize cardinals, and in response cardinals, like some other birds, counter by "anting." They find an anthill, pick up the insects in their beaks one by one, and rub the crushed bodies all through their plumage. Since ants contain high levels of formic acid, ornithologists have long supposed this to be a means of discouraging parasites. No one seems to know for sure. In any case, a bird that's just finished a thorough round of anting looks as wet as one that's had a dip in a rain puddle.

◆

Clearly, none of the hazards, neither individually nor in concert, has any significant effect upon cardinal populations. As with all wildlife, habitat is the key, and new cardinal habitat has been a byproduct of human activity since colonial days. Concrete wilderness isn't much to their liking, nor is a monocultured landscape where nothing grows taller than a stalk of wheat, but *Cardinalis cardinalis* has as readily taken to suburbia as it did to land-use changes a hundred years ago. No doubt the species continues to expand its range even today.

Which is a pleasant prospect for anyone who enjoys sharing living space with a diversity of birds, and pleasanter still for those with an ear for song. Less versatile than the mockingbird and perhaps not so virtuosic as the warblers, cardinals are nonetheless singers of infinite charm. They are, moreover, among only a handful of birds inclined to sing year-round, although they naturally are most persistent in the spring, especially in the northern parts of their range.

Amelia Lasker, who made an important systematic study of cardinals in Tennessee during the 1930s and early '40s, documented at least twenty-eight different songs, all of them sung with equal fervor and skill by both sexes, though the females' vocalizing not surprisingly subsides during the nesting season. Cardinals like to sing from treetops, aren't averse to singing in flight, and sometimes even sing in the dark if disturbed on their nighttime roosts. Perhaps that's why in some parts of the Southeast they are locally called Virginia nightingales.

Among those who enjoy translating bird songs into English, parts of the cardinal's repertoire have been rendered as *peer, whoit, cheedle, pheu, phey, teeyo, toowee, to, wheeteeyo, whitowee, whitcheeah, toolit, tayo, cue, hip-ip-ip, e-eee, weet, we-oo, birdy, tik, chuck-er whee,* and some others as involuted as an auditory Rorschach test.

Perhaps the most familiar cardinal song, and the one that falls most sweetly on my ear, is a little melody that sounds to me something like *whit...whit...whit...whit-cheer...whit-cheer...purty-*

*purty-purty...whit-cheer*. You can hear it virtually any time of year, but it seems a particular favorite in the spring. The woods where I first heard it are gone, and in any case, I now live far away. My nighttime serenades now come from whippoorwills instead of screech owls, but cardinal songs still measure out my days, still take me back to the house on the edge of town and to a time when all the world seemed new.

◆

# 20

# Fragments of the Rainbow

Until the autumn of 1492 when Christopher Columbus made landfall in the West Indies, no European had ever seen a canoe, tobacco, or a hummingbird. Impressed with canoes and tobacco, the voyagers readily adopted both items and also the Arawak Indian words for them. But if they paid any heed to hummingbirds they left no record.

It is a rare man who can remain unmoved and unenthralled by a hummingbird, although in his defense, the great Admiral of the Ocean Sea was preoccupied at the time with trying to figure out where in the world he was. Besides, neither Columbus nor his companions had any way of knowing that the tiny creatures buzzing insectlike among the island flowers represented one of the largest families of birds in the world, nor that every member possesses fantastic gifts unique to its kind.

Not even the truly knowledgeable men of science who came later could fully appreciate what they saw. But none remained long in the New World without recognizing in hummingbirds some special touch of nature's hand—as did their Old World colleagues once dried skins and living specimens reached Europe. Clearly charmed by their size and brilliantly iridescent plumage, normally staid naturalists were moved to coin fantastic names reminiscent of precious gems and fairy tales: crimson topaz, horned sungem, blue-chinned sapphire, Natterer's emerald, Brazilian ruby, blossomcrown, adorable coquette, black-hooded sunbeam, violet-capped woodnymph, purple-crowned fairy, glowing puffleg, violet-tailed sylph, blue-throated star-frontlet, and dozens more.

Some even translated their lyrical fancy to Latin and Greek, creating such generic names as *Stellula*, "little star"; *Chrysolampis*, "golden torch"; *Sapphirina*, "sapphire." Others went whole hog, naming species *Topaza pyra*, "fiery topaz"; *Heliomaster furcifer*, "seeker of the sun"; and loveliest of all, *Calothorax lucifer*—"beautiful-breasted light-bearer."

Central and South American Indians know hummingbirds by names that translate as "rays of the sun" and "tresses of the day-star." To some Spanish-speakers of South America they are *picaflor*, "flower-pecker," to Brazilians *beijaflor*, "flower-kisser."

The French, for once uncharacteristically dull, call them *oiseau-mouche*, "fly-bird," but eighteenth-century French naturalist Georges Louis Leclerc, le Compte de Buffon, summed up everyone's feeling for hummingbirds by declaring them "of all animated beings...the most elegant in form and brilliant in color," bestowed by nature with "all the gifts of which she has only given other birds a share."

It's hard to disagree with Monsieur le Compte and equally difficult not to be astonished, as he was, by nature's skill in fitting such gifts into exquisitely tiny packages. Not even the largest of them—the eight-inch giant hummingbird of the Andes—weighs an ounce. The smallest, the bee hummer of Cuba, is the smallest of all birds, indeed the tiniest warm-blooded animal; it's about $2\frac{1}{4}$ inches long, half of which is beak and tailfeathers, and weighs less than a half-dozen aspirin tablets.

Between the giant and the bee are 318 other species making up the family Trochilidae, the second-largest family of birds in the New World. (With 367 species, the American flycatchers are the largest.)

Although hummers have long since been transplanted elsewhere, they occur naturally only in the Western Hemisphere. By far the majority live in the tropics of South America within a band extending five degrees on either side of the equator. From there in both directions the number of species diminishes, although hummers are able to live at least part of the year virtually throughout the Western world with one species, the rufous hummingbird, reaching as far north as Alaska and another, the green-backed firecrown, as far south as the Strait of Magellan and Tierra del Fuego.

More than fifty species occur north of Panama, but only about fifteen show up here, half of which get no farther north than the tip of Florida or barely across the Mexican border. Only eight species occupy any significant range in the United States: Anna's, calliope, broad-tailed, black-chinned, Costa's, rufous, and Allen's hummingbirds in the West and a single species, the ruby-throat, east of the Great Plains.

Although most of the world's hummers live in the warm equatorial belt, they're hardy little specimens. Dozens of species live in the high Andes of Colombia, Peru, and Ecuador where nighttime temperatures fall to the freezing point. Some get along quite happily at altitudes of nearly 15,000 feet, just below the alpine fields of perennial snow. Wherever you find flowers in bloom, you're likely to find hummingbirds.

Like the gods of Olympus, hummers are great drinkers of nectar. They supplement their liquid diet to some extent with tiny insects and spiders, but nectar is their thing. During millions of years spent exploiting this feeding niche, they have evolved some adaptations even more extraordinary than the old-time naturalists, breathless with wonder, could have guessed.

◆

Apart from a few insects, no flying thing comes even close to matching the sophistication and consummate skill of the hummingbird. Any hummer, regardless of species, can fly forward, backward, up, down, or hover motionless in midair. It can pivot on the axis of its own body, fly sideways, do an extended backward roll upside down, and dart from here to there in milliseconds. A hummingbird can do anything on the wing except soar. A peregrine in full stoop is faster, but a hummer can come close to a mile a minute in a courtship power dive and can reach nearly thirty miles per hour without being in any great hurry.

Such capabilities call for specialized mechanics, and hummingbirds' flight equipment is indeed unique. In the structure of its bones a bird's wing is not unlike a human arm. Each has three joints—shoulder, elbow, and wrist—and in a typical wing the bones are similar in proportion to ours, with upper and forearm bones of about equal length and the remaining limb relatively short from the wrist on out. Like those of the other great fliers, the swifts, hummingbirds' wings are proportioned just the opposite way: the upper bones, from shoulder to elbow and elbow to wrist, are quite short and the lower ones, forming the "hand" portion, quite long. In hummingbirds the upper bones are fixed in a fairly rigid V-shape, and neither the elbow nor wrist joints permit much flexure. The shoulder joint, however, is extremely supple, so the wing can move in all directions and can even rotate a full half-circle on its own axis.

Our own shoulders are in some ways equally limber, but to see where hummingbirds go us one better, extend your arm straight out with your palm down, then turn your hand palm-up. The rotation, you'll notice, is mostly in your forearm; in a hummingbird's wing it's all in the shoulder. Add to this flexibility some highly specialized musculature, and you have the key elements in a hummer's remarkable abilities of flight.

Birds fly using their breast muscles, two sets on each side, supported by a keel of bone down the center. The smaller elevator muscles, lying close to the breastbone, allow a bird to raise its wings, and the depressors—the larger outer muscles—supply the down-stroke. Because the down-stroke is for most birds the sole source of aerodynamic lift and propulsion, the depressors typically are ten to twenty times larger than the elevators. This is not so of hummingbirds.

Proportionate to body size, hummingbird breastbones have the deepest keels of any bird on earth. Their flight muscles are also larger, making up as much as 30 percent of total body weight, and the elevator muscles are half the size of the depressors. Thus equipped, a hummer derives lift and propulsion from both upward and downward strokes just by altering the angle of its wings. To hover, the bird beats its wings forward and back rather than up and down, and rotates them at the shoulders on each stroke so that the front edge is always the leading edge. This creates a perfect equilibrium of opposing forces and holds its body precisely in one spot.

Hummers use their exquisite skills to the fullest. Their legs aren't much good for anything other than perching. An incubating female, for instance, flies directly onto her nest rather than landing on the rim and hopping in as other birds do. If she wants to change position, she hovers up an inch or so, turns and settles down again. Hummers even bathe on the wing, taking a flying dip into a pool, fluttering through dewy leaves and even small waterfalls.

Most species' wingbeats are rapid enough to create the faint buzz that prompts us to call them hummingbirds. In proportion to body size, hummers actually beat their wings more slowly than other birds, largely because their flight apparatus is more efficient. Moreover, hummingbird wingbeats vary according to the size of the bird and what it's up to at the moment. In normal flight the giant hummer strokes its wings about eight times per second while the minuscule white-bellied woodstar might make seventy-five beats per second. The ruby-throat and others of similar size average about fifty-five strokes per second while hovering.

◆

Their ability to hover is one chief reason why hummingbirds are able to exploit nectar as a food source. Their bills and tongues are another. Although a great many species, including those most common in North America, wear long, fairly straight, needle-shaped bills, the tropical species show a tremendous diversity—from the sword-billed hummer whose four-inch beak is nearly as long as its body and tail together, to the purple-backed thornbill whose 5/16-inch beak is shortest of all.

Between these extremes are bills of every size and shape. Some are straight, others curve gently, some are sharply bent, and those of a couple of species even curve upward. In general, the length and configuration of its beak defines each species' sub-niche for feeding: The longer the bill, the deeper the blossom the bird can feed from, although at least one short-billed species, the purple-crowned fairy, literally taps into its food source by piercing petals at the base of a blossom.

In any case the bill's main function is to protect a hummer's tongue, which is the real key to its livelihood. Some are extremely long—longer in fact than the bill itself—and some are considerably shorter, but all hummers' tongues are wonderfully specialized tools. Essentially, the organ comprises two parallel tubes arranged side by side, separated toward the tip and fringed at the ends. Hovering before a blossom, the hummer inserts its bill and extends its tongue to the nectar inside. Whether it then sucks the liquid in as we might drink through a straw or whether capillary action plays a part, no one knows, but a hummer does not lap up its food like a cat, as was once supposed.

Nectar contains a great deal of natural sugar with an extremely high energy content. Hummingbirds are voracious feeders even so, taking in about half their own weight in sugar every day. They have to, because a hummer is a dynamo in feathers, expending more energy per unit of weight than any other warm-blooded animal.

As part of a landmark study during the 1950s Crawford Greenewalt extrapolated some comparisons between hummingbirds' energy output and our own. In normal activity a 170-pound man expends about 3,500 calories every day. To perform at the level of a hummer going about its daily round the same man would have to burn about 155,000 calories, and he would have to eat 285 pounds of ground beef to supply the fuel. A hovering hummer, Greenewalt says, expends ten times more energy than a man running nine miles per hour. If a man could perform at the hummer's level—if he could, in other words, run ninety miles per hour—he would have to evaporate about a hundred pounds of sweat every hour just to keep his skin temperature below the boiling point of water. If he stopped sweating altogether his skin temperature would rise to 750 degrees Fahrenheit, at which point he would burst into flame.

Given their strictly diurnal, high-intensity lifestyle, nighttime poses a problem for hummingbirds, especially for those that live in cool climates—whether in latitudes far from the equator or at high altitude. Some Andean species avoid extreme cold by roosting in caves. In normal sleep a hummer's body temperature may decline five or six degrees, which is some help in conserving energy. But when greater economy is called for the bird can go into overnight hibernation—noctivation, as one researcher puts it. In this condition the bird's body temperature plunges almost to the temperature of the air around it, breathing slows, and its total energy output is reduced to about one-twentieth that of normal sleep. In hummingbird terms this amounts virtually to suspended animation—though a torpid hummer still uses as much equivalent energy as a man walking briskly.

Only the birds themselves know exactly what conditions prompt the shutdown. How well-fed a hummer happens to be at nightfall probably is one factor. Ambient temperature appears to be another, since some observers have noticed that birds do not grow torpid on nights warmer than about ninety-three degrees nor when the difference between nighttime and daytime temperatures is less than twelve degrees. To compound the mystery, some hummers noctivate every night while others don't—even birds of the same species roosting side by side. A female incubating eggs or brooding young doesn't noctivate at all, nor do her older nestlings even if she leaves them exposed.

Which well she might, because a pair of nearly full-grown hummers is more than a nestful. Besides, by the time they're that far along she may feel in need of a break as she has reared her young all by herself.

A male hummer is a travelin' man dressed in a suit of lights so resplendent that Aztec nobles wore cloaks made from hummingbird skins. The Indians of eastern North America made earrings from ruby-throats dried whole. Audubon described hummers as "glittering fragments of the rainbow." To recall Buffon's comment, nature gives turkeys, pigeons, starlings, and crows at least a share of iridescent plumage; to hummingbirds she gives it all, every hue of the spectrum, and naturalists wrangled for years over both the nature and purpose of all their brilliant colors.

Iridescence is entirely a function of reflected light, not of pigment. Inside the barbules of iridescent feathers lies a mosaic of minute, elliptical platelets filled with air bubbles, visible only through an electron microscope. Light striking these platelets is shattered, the same way light is shattered by soap bubbles or a thin film of oil on the surface of water. Instead of reflecting as a single color the light is therefore broken into many colors. The platelets' differing shapes, thickness, and air content determine which hues predominate.

How we perceive iridescence depends upon the light source and our angle of view. The greens and blues of a hummingbird's back look much the same from any angle since the barbules of those feathers are curved like tiny concave mirrors that reflect in many directions at once. But if he turns only a few degrees in relation to the light source and our eyes, the flat barbules of his gorget or crown might appear red or cinnamon-orange, violet, blue, green, or black, all in turn.

The function of such shimmering gorgeousness is considerably simpler, at least so far as the bird is concerned; it's his ticket to one-night stands with as many females as he can tempt. Courtship is mainly a male affair, although the fact that females often build nests before making themselves available for wooing makes me wonder who's really in charge of things.

In any case, males make the first overtures, which for many tropical species involves singing, often in groups of a hundred or more, performed at some traditional singing ground. Even in chorus it is largely *sotto voce* work since none can sing loudly. In fact, few can "sing" at all, at least by comparison with most other small birds. A tired old joke has it that hummingbirds hum because they don't know the words, but the old chestnut has some truth in it. Only a few species are capable of anything more melodic than some monotonous, rusty-sounding creaks, chirps, croaks, ticks, and twitters.

It is, however, enough, and presently the females show up. Not many are as brightly colored as the males, and the greater the differences between his plumage and hers, the more elaborate his courtship performance. Some, like the racket-tails of South America, augment their iridescent finery with a pair of specialized tailfeathers—bare shafts two or three times longer than the bird himself, with fan- or paddle-shaped ends. At least one species, the aptly named marvelous hummingbird of Brazil, has such fine control over these feathers that he can raise them one at a time like semaphores while hovering in front of his lady.

For the most part the males' displays are extravaganzas of flight and color, involving long, looping, intricate, high-speed aerobatics with occasional pauses to hover in front of the females and flash all their dazzling colors. Females often signal their willingness by joining the males in courtship flights that would visit humility upon the head of any Olympic figure skater.

No matter how complex the preliminaries, the resulting relationship is fleeting, and without so much as a cigarette or a by-your-leave he's off in search of new conquests. Besides, his erstwhile paramour has an agenda of her own.

◆

The nature of her nest varies according to species. It may be a simple cup anchored to a tree branch or, for the hermit hummer of the tropics, an elaborate hanging structure fastened to the side of a living leaf and counterweighted with pellets of mud so it doesn't tip under the combined weight of nest, bird, eggs, and young.

All species common to the United States build cup-type nests, but you can spend a long time looking before you find one, even where the birds are abundant. It's a minute affair scarcely two inches across, made of seed down, moss, woolly leaf fibers, lichens and such, wound with spider silk and usually so well blended to the surroundings that it looks like part of the branch it's on.

The giant hummer often lays a single egg, but all the rest lay two—long, narrow, pure white, and wonderfully tiny. When they appear after about two weeks' incubation the hatchlings are tinier still, smaller than a fingernail, blind, naked, utterly helpless to do anything but gape their bills for food.

Their only talent at that age is sword swallowing, and they get plenty of practice. Every few minutes the female buzzes to the nest, pushes her needlelike bill down their little gullets, and literally pumps them full of regurgitated nectar. It's a hair-raising thing to watch, because you're sure she's going to fly off with one of the kids impaled on her beak.

It never happens though, and the young ones grow rapidly—although the time they spend in the nest varies a great deal, probably according to the amount and quality of food the female is able to find. Young ruby-throats have been known to leave the nest after only ten days or stay as long as a month. In any event it's a harrowing time for the female, who must satisfy not only her own requirements for food but those of the nestlings as well. Since she doesn't economize her energy through noctivation during this time, one can only imagine that she's thoroughly frazzled by the time the next generation takes off on its own.

◆

She certainly is short-tempered. Hummers in general are feisty, fearless little things, continually squabbling among themselves over feeding territories. At nesting time they're willing to attack anything that might be a threat, regardless of size—songbirds, hawks, owls, squirrels, housecats, you name it. I once watched a female ruby-throat try her best to pick a fight with my old farm dog, who weighed better than seventy pounds. Waldo didn't exactly flee the scene, but he didn't stick around long, either.

At other times their audacity seems more a matter of curiosity, for hummers appear endlessly interested in everything around them. It probably is their way of finding new sources of food, and it certainly offers us the opportunity to see them up close. I keep feeders on my patio, and that's my favorite summertime reading spot, partly because every so often I'll hear wings and look up

to find a hummer hovering a few inches in front of my face, perhaps wondering if there's a blossom anywhere in that foliage of beard. If I sit still and hold out some bright-red object, one occasionally will perch on my hand for a moment and check it out.

For all their physical intensity hummingbirds can live a surprisingly long time, ten years or more, but like other small birds most of them don't. Nesting losses are high. Predation claims some, weather and starvation probably more. A fair number fall victim to an array of accidents—from flying against windows and glass doors to entangling themselves in spider webs.

In fact the first hummingbird I ever saw at really close range was a male ruby-throat I found trapped in spider silk, high up at a gable window in my grandfather's garage. I was nine or ten at the time and with some help from my father managed to get the little chap down, still alive. Once we got the worst of the cobwebs off, it perched quietly on a twig I held between thumb and finger while Dad worked with toothpicks and swabs to remove as many of the remaining sticky strands as he could. We offered a feeder and a solution of honey and water, but it apparently was too exhausted to drink and died an hour or so later.

Our friend Mr. Ayres, the bird man, showed me how to make the tiny carcass into a study skin, which I kept for years. I don't believe I ever looked at it without feeling a profound sense of wonder at its miniature perfection. The specimen is long since gone, but the wonder of it is with me still.

# V

# Warmth and Wonder

# 21

# Warm Blood, Cold Seas

Of all the environments on earth none is more alien to us than the ocean depths. The cold, dark, liquid world below the surface offers no oxygen to our air-breathing lungs, exerts more pressure than our frail bodies can stand, and leaches away body heat far faster than we can generate or preserve it. Even if we manage to avoid drowning or hypothermia we soon die of thirst, because we simply cannot stomach seawater's quotient of salt.

Hard to imagine a more hostile environment for a mammal. It's one thing to live along the margins of the sea, ride its surface, and even probe a fraction of its depth. Millions of mammals do, including man himself. To live an entire lifetime in the water, however, feeding and breeding and rearing young, is something else again—and that makes the sea-living mammals extraordinary indeed.

Among the questions accruing to the notion of warm-blooded, air-breathing, gestating, birth-giving animals living aquatic lives, the least mysterious is *why.* In the 600 million years since multicelled organisms first appeared, the oceans have become enormously rich in life forms, from the microscopic to the immense, plant and animal alike. These abundant resources of food are available to any animal with the means of exploiting them, and it was this, ages ago, that attracted certain mammals to the sea.

The process was something of a homecoming, a journey back to the origins of life itself, for none of the marine mammals actually evolved in the sea directly from reptilian ancestors, but rather on land from other mammals millions of years removed from life in the water. Like other repatriate forms—penguins and sea turtles and such—they represent an astonishing convolution in the development of life on earth: Having achieved the most radical transition possible, from aquatic life to terrestrial life, they readapted to their ancestral environment while still preserving all the characteristics they'd developed in order to live on land. Marine mammals did not, in other words, re-evolve into reptiles or amphibians but instead remained mammals—a bravura performance of adaptability.

Such transition could hardly happen overnight. The cetaceans—the whales, dolphins, and porpoises—have been at it longest, followed closely by the sirenians—the manatees, dugong, and sea cow. Both groups evolved from a primitive hoofed mammal that waded shallow bays and estuaries foraging for shellfish. The cetaceans probably took to the sea about 60 million years ago, the sirenians perhaps 5 million years later.

The pinnipeds—the seals, sea lions, and walruses—appeared about 30 million years ago, descended from two different ancestors, both carnivores. According to the fossil record one was bearlike and gave rise to the modern group known as walking seals, those whose flippers can still serve as legs when the animal is ashore. The crawling seals, whose hind legs have become so specialized for swimming that they cannot support their bodies, derive from an otterlike carnivore.

◆

How long they've been about the task correlates closely with the degrees to which the various groups have adapted to life in the sea. The cetaceans, most specialized of all, never leave the water in the course of normal life. Sirenians occasionally wallow onto reefs to bask in the sun. The pinnipeds, still somewhat linked to the land, must clamber ashore or onto ice floes in order to mate and give birth.

Nonetheless, all marine mammals face essentially the same problems in finding a livelihood and reproducing in a cold, wet, salty, three-dimensional world.

Adapting to salt water seems to be the least troublesome. Lots of animals that live in salt marshes thrive perfectly well on seawater. No one fully understands the physical mechanisms by which marine mammals keep their bodies hydrated. Certainly they swallow some quantity of water involuntarily, and like desert animals that seldom if ever actually drink, they manufacture a considerable amount of metabolic water from their body fat. Some cetaceans—the bottlenose dolphin, for instance—have unusual mucous glands on their tongues that may help remove excess salt.

At any rate, salty water isn't a problem even for those species that customarily trade between freshwater rivers and the ocean. Harbor seals, manatees, the beluga, and some species of African, Asian, and South American porpoises are equally comfortable in fresh or salt water, and their kidneys rapidly adjust to differences in salinity.

As a medium, however, the water itself presents complex challenges. It is denser than air and therefore poses more resistance to anything moving through it. Water also conducts heat efficiently, about twenty times better than air. This is a critical problem for warm-blooded animals that must maintain core body temperature at a relatively high level and within limits of only a few degrees.

As the marine mammals adapted to aquatic life, their bodies evolved toward ever more streamlined shapes. Legs shortened and became flippers or disappeared altogether. Some appendages—

nipples, sexual organs, and the like—withdrew into pockets or pouches in the skin. External ear flaps either grew tiny, as in the walking seals (which are sometimes also called the eared seals), or disappeared. The cetaceans' neck bones slowly telescoped to form sleeker, more rigid bodies.

Although useful for insulation, hair creates drag in the water. The cetaceans have lost theirs entirely, including eyelashes, although all species sport a few hairs while in embryonic stage, and some whales have a few adult bristles around their mouths. Sirenians, too, are nearly hairless except for a few vibrissae, or tactile hairs, on their lips. Seals have not yet lost their hairy coats, which reflects their relatively shorter evolutionary span. Besides the suit of guard hairs that all pinnipeds wear, two genera of walking seals also grow layers of underfur. These coats are extremely dense, as many as 300,000 hairs per square inch, and trap minute air bubbles so efficiently that healthy fur seals never get wet clear to the skin.

In keeping with the trade-off their pelage exacts, pinnipeds are highly maneuverable but not particularly powerful swimmers. A sea lion can get up enough speed to clear a hurdle seven feet out of the water, and an escaping fur seal might make ten miles per hour, but even the ponderous, sluggish manatee can swim half again faster than that.

As a group the cetaceans are the finest swimmers of all. A cruising blue whale develops about forty-five horsepower, and a gray whale, which might weigh twenty-five tons, is powerful enough to hurl its body completely out of the water. Porpoises of several species can top twenty-five miles per hour; killer whales, which actually are a species of dolphin and are the nonpareil of swimmers, have been clocked at nearly thirty-five.

As they lost the insulation provided by a coat of hair the marine mammals developed other ways of preserving body heat. All of them, seals included, are enveloped in a layer of blubber, which is composed of fat and oil and is laced with blood vessels.

A ringed seal's blubber might make up half its total body weight, that of a walrus or elephant seal perhaps a third. In a whale the layer may be two feet thick.

Their circulatory systems, too, are specially adapted for conserving heat. At the extremities, arteries and veins lie close together so any heat that escapes from the warm arterial blood is captured by the veins and kept inside the animal's body. Seals and smaller cetaceans generate additional heat through a metabolic rate higher than that of most terrestrial mammals. In the big whales, however, metabolism carries on at a remarkably slow pace; no one has yet identified the exact rate, but it's believed to be about half that of a sloth—and some species of plants move faster than sloths.

The difference has much to do with body size, which is another characteristic of marine mammals' splendid adaptation to the sea. Small objects lose heat faster than large ones, and no mammal that weighs less than about 15 pounds can live full-time in the open ocean. On the other hand, seawater is particularly buoyant, so freed from the necessity of supporting their own weight, marine mammals can grow to enormous size.

Depending upon species, adult seals and sea lions range from about 80 to more than 8,000 pounds. Sirenians weigh about a ton on the average, but the Steller sea cow—the only sirenian to have lived in Arctic waters—was about 25 feet long and weighed upwards of 7 tons. It has been extinct since 1768.

The cetaceans come in the widest range of sizes, from 50-pound dolphins to the blue whale, the largest animal ever to live on earth. A mature blue whale might be 90 feet long and weigh more than 175 tons, roughly the equivalent of thirty-three elephants.

Besides helping the animal maintain a stable core temperature, size also is an advantage in withstanding the enormous pressure of deep water. And some marine mammals go deep indeed. A Weddell seal has been tracked by instruments at nearly

2,000 feet below the surface. The bodies of whales have been discovered tangled in underwater cables even deeper yet—a killer whale at more than 3,300 feet and a sperm whale at 3,700 feet.

Diving to even a fraction of those depths and remaining there for an hour or more presents some difficult problems to an animal that must depend solely upon oxygen stored in its body. None of the marine mammals have particularly large lungs in proportion to body size, but they have evolved other, more efficient means of caching an oxygen supply.

Blood and muscle tissue are the primary reservoirs. Seals' bodies contain great quantities of blood, perhaps twice as much as ours in proportion to body size. Cetaceans' muscles contain high levels of myoglobin, a protein particularly efficient at storing oxygen.

A marine mammal also uses its oxygen economically. When it dives, its heart rate subsides by as much as 90 percent, metabolism slows, and its circulatory system reduces the amount of blood going to the extremities and shunts more blood to the heart and brain. Moreover, the respiratory center of the brain tolerates relatively high accumulations of carbon dioxide, and the nervous system has a sort of automatic shut-off that prevents the animal from attempting unconscious breathing, even if it's momentarily stunned underwater. Since marine mammals' nostrils are flap-shaped and held shut by water pressure, it would be difficult if not impossible for one to draw a breath at any great depth even if it tried. These animals certainly can suffocate, but they rarely drown.

Humans breathing compressed air underwater are liable to a painful and sometimes fatal affliction called the bends. Under pressure, nitrogen in the air dissolves into body fluids and tissues and then, as the pressure lessens when the diver ascends, comes out of solution as bubbles. Human divers must therefore surface in stages, allowing time for the nitrogen bubbles to dissolve. Marine mammals have no such difficulty. Since they operate underwater on only a single lungful of air at a time, they aren't breathing air under pressure, and their bodies need deal with only a small amount of nitrogen. Nitrogen doesn't get into their blood or tissues anyway; in a deep dive the animal's lungs compress, and the air in them is driven into its windpipe and nasal passages, which are lined with a thick membrane impervious to gases.

When it surfaces the animal simply exhales, draws in a breath of fresh air, and can dive again immediately. Among cetaceans, exhaling can be a spectacular affair. In all species, the nostrils have migrated to the top of the head. These are universally denoted by the less-than-delicate term *blowhole,* and exhaling is commonly called *blowing.* And an impressive blow it is, especially in the big whales, a breathing-out so forceful that the water around the blowhole is blasted upward in a plume of vapor visible for miles on the open sea. Old-time whalers believed the animals were exhaling streams of water, and blowing still is often called *spouting.*

Besides their specialized mechanics, marine mammals have developed sensory adaptations as well. Of the five senses, smell and taste are least useful and are progressively duller according to how far each group is removed from terrestrial life. Seals have fairly acute senses of smell and taste, sirenians only poor ones, and cetaceans almost none at all.

Capabilities of sight are similarly varied. Except for the walruses', all seals' eyes are large, those of sirenians and cetaceans small. All of them can see better underwater than in the air, and their eyes are well adapted for gathering available light. Even so, touch and hearing are far more useful senses.

Water is an excellent conductor of sound waves, and hearing is highly developed among all marine mammals. Indeed, the porpoises, dolphins, and some whales use echolocation, just as bats do. They produce short, click-like, ultrasonic sounds in their nasal passages and focus them by means of a fatty, lens-shaped body called a *melon,* located in the forehead. Sound waves reflecting back from nearby objects channel through oil-filled cavities in the lower jaw, directly to the inner ear. This apparatus is so sensitive that porpoises can discriminate the texture of potential prey by sound alone.

But it isn't perfect. The mysterious instances when cetaceans beach themselves with what appears to be almost suicidal determination may actually be cases of sonar systems gone awry or may result from the system's inability to recognize a gently sloping beach. In any event the species in whom echolocation is least developed rarely end up stranded.

Marine mammals are as adept at making sounds as they are at hearing them. From seals to the great whales, they give voice to an impressive array of barks, roars, moans, chirps, screams, whines, coos, snarls, hisses, whistles, grunts, and groans. The younger the animal in evolutionary terms, the more likely it is to vocalize out of water. The pinnipeds that breed socially in vast rookeries are noisiest of all.

The cetaceans, who seldom make sounds out of water unless trained to do so, are the most intriguing, for they communicate by voice in highly sophisticated ways, using rich vocabularies. Researchers have intensively studied whale and porpoise "languages" in recent years, and while we're a long way from any thorough-going conclusions, we've learned some astonishing things. We know for example that the animals use words—in the sense, of course, that a word simply is an arbitrary sound that through custom has acquired certain meaning. We know also that dialectical differences occur among isolated populations in a given species and that some interspecies communication takes place among porpoises.

Being big animals, whales have big voices that may carry for a hundred miles or more through the sea. Whale songs, elaborate constructions of whistles and creaks and moans, are to my ear among the most haunting, emotionally moving sounds in nature. It's a pity more of us can't hear them firsthand more often, but a few recordings are available. They'll make you think about all sorts of things. And if by some means we ever are able to communicate directly with the whales, large or small, who knows what fabulous stories they might tell?

Like terrestrial mammals, those that live in the sea exploit their environment for a livelihood in a variety of ways. All of the pinnipeds are carnivores, feeding on fish, squid, mollusks, crabs, shrimp, sea urchins, starfish, penguins, gulls, and other pinnipeds. Some species forage along the bottom; others hunt the upper waters.

The sirenians are strictly vegetarian, endlessly browsing and grazing on sea grasses and flowering aquatic plants in warm, shallow waters.

Perhaps the more diverse because of their longer residence in the sea, the cetaceans fall into two groups according to their feeding habits. The Odontoceti, or toothed cetaceans, comprise all of the porpoises and dolphins and the white, sperm, and beaked whales. These, as the name implies, grow sets of teeth and are carnivorous. Fish, squid, octopi, and crustaceans are favorite prey. The killer whale is the only cetacean that makes a habit of hunting warm-blooded animals as well—seals, sea lions, porpoises, and even small whales. Killers working in packs even attack big whales but apparently only if the larger animal is injured or trapped in a shallow tidal basin.

The largest whales, curiously, survive on the smallest prey. The Mysticeti, or baleen whales, evolved the wherewithal to harvest the richest of all the food sources in the sea. From transverse ridges in the palates at the roofs of their mouths, two rows of thin, bristly plates hang on each side. This substance, called baleen and once prized as "whalebone," actually is modified mucous membrane, and it serves as a filter through which the animal strains minute organisms from the water.

Zooplankton is a motley collection of life forms in hundreds of species—tiny crustaceans, plants, jellyfishes, sea worms, and other stuff collectively called *krill*. No one knows how much krill the world's oceans contain, but it forms the base of virtually the entire marine food chain. It also clearly represents a vast resource of energy, for the baleen whales and the largest species of shark thrive on it. Crabeater and ringed seals also eat a great deal of krill in addition to their other foods; not surprisingly they are the two most numerous species of seals on earth.

Baleen whales feed simply by swimming through zooplankton swarms with their mouths agape and lifting their tongues to squeeze out the water, leaving the krill trapped in the baleen. They may thus consume as much as two tons per day.

Other whales occupy different feeding niches, and their baleen is adapted accordingly. Gray whales scour and forage along the sea bottom for relatively large organisms; their baleen therefore grows relatively short and stiff. Baleen grows longest, up to twelve feet, and more finely bristled in bowhead and right whales, which allows them to capture the smallest zooplankton.

Basing their arguments upon the distinctive physical differences between toothed and baleen whales, some zoologists insist that they evolved from different ancestors. More likely, however, the Mysticeti evolved from the Odontoceti, and the most compelling evidence is the fact that most species of baleen whales have teeth in the earliest stages of fetal development.

Reproduction poses relatively few problems for marine mammals, although those that never leave the water have developed some special adaptations. Except among the migratory whales, mating is less sharply seasonal than among land mammals. None of the marine species bear young more frequently than once per year, and the larger cetaceans give birth only once every two to five years. For killer whales the interval can be as long as eight years. Single births are the general rule for all marine mammals.

Gestation may take a long time, up to seventeen months in the big whales, but it's no special problem, since nearly every animal in the world spends its embryonic and fetal life in water—pond, sea, river, lake, or the miniature ocean inside an eggshell or a womb filled with amniotic fluid. At the start we all share something of our most ancient ancestors. On the other hand, to be an air-breathing animal born in water is something of a challenge. The typical headfirst mammalian birthing posture clearly would be a hazard, so cetaceans and sirenians are born tail-first. Pinnipeds, as I mentioned earlier, are born on land or an ice floe.

All marine mammals are precocious at birth, but only those actually born in the water must swim immediately. In some species of whales several adults may act as midwives, helping newborns to the surface and supporting their bodies while they take their first breaths.

Seals wean and abandon their young relatively quickly—in less than three weeks for some species, in about four months for others. Walruses, some sea lions, and all cetaceans nurse considerably longer, some species for a year or more. In any case, all young marine mammals grow rapidly, fed on milk that contains about 50 percent fat. Since cetaceans have no lips with which to actually suckle, the females expel their milk forcibly and the young funnel it down with their tongues. It works well enough, and the milk is rich enough, that over its six-month nursing period a young blue whale grows at the rate of nearly nine pounds per hour.

The marine mammals demonstrate some poignant truths about the nature of life on this planet. For one thing they show that adaptability has remarkably few limits. They also offer graphic evidence that evolution has no particular end but rather is a continuous process pointing always toward some future, unwinding at levels of time beyond our ability to comprehend. For illustration we need only look at two animals in the process of becoming marine mammals.

Some mammalogists consider the sea otter already more marine than terrestrial. Its body still has the essential, four-legged shape of a land animal, although its hind feet are flipperlike and it spends most of its time in water. Its underfur is even denser than seals', 600,000 hairs per square inch. It floats on its back between dives and in fact can swim faster in that posture than it can belly-down. Like the pinnipeds, sea otters breed and give birth on land, but only about half the young are born headfirst. Fossil evidence suggests that sea otters have lived in their present form for about 5 million years.

Although by all definitions a terrestrial animal, the polar bear clearly seems on an evolutionary path leading toward life in the sea. It spends most of its life swimming or riding ice floes. It has good underwater vision, can close its nostrils when submerged, and can stay under for several minutes at a time.

A definitive answer to whether these animals will eventually become as thoroughly ocean-going as the whales lies far off, perhaps as long into the future as the origin of the whales lies in the past. But unless their evolutionary progress is disrupted, they probably will, given world enough and time.

◆

# 22

# Water Sprites

We were three tired grouse hunters watching shadows gather under the Colorado hills, easing the not-unpleasant aches of a day's walk with sips of whisky and a fine long view toward the Williams Fork Mountains. As hunters' talk often does, ours rambled among various animals we're especially fond of, and Bob concluded some comment by saying, "But the ones most fun of all to watch are..."

Without a moment's prior thought, as if on cue, Jon and I joined in and we all spoke the same word in perfect synchrony.

"...otters."

Such things happen sometimes, especially among old friends—to say nothing of how great minds work—but I can see some reasons in this one, and some telling truths. For one thing we were all boys at the same time, in the late 1940s and early '50s; those were good years in which to be boys, especially boys in love with wild things and wild places, because there was more of both available then than at any time since.

For each of us those wild places were full of rivers and streams. Jon had the Mississippi in eastern Ontario, Bob the Brazos and Trinity rivers of Texas, and I the Des Moines and South Skunk in Iowa. Even such far-flung geography tells a certain tale; that three boys such distances apart could be captivated by the same animal indicates a species whose range covers a lot of country. But even more, it points out the bottom-line fact of the matter: Spend any time at all in the company of otters and you'll love them for the rest of your life.

Actually the distance between eastern Canada and the Texas Gulf is a mere fraction of the territory otters occupy, for they're a goodly clan comprising a dozen species in six genera, and in one form or another they live on every continent but Australia. Africa is home to the spot-necked and Cape clawless otters; Asia has Indian smooth-coated, Oriental short-clawed, and hairy-nosed otters, and shares some European river otters as well. South America has more species than any other continent—giant, marine, southern river, and Neotropical river otters. In North America we have two, only one of which is really widespread. *Enhydra lutris*, the sea otter, is the rarer of the two, and lives at certain places around the Pacific rim, most notably along the coasts of Alaska and southern California. The one we aging grouse hunters had in mind is *Lutra canadensis*, the river otter—the finest fisherman on legs, an amiable creature and a raffish clown who seems as fond of frolic as any animal alive.

I don't know how Bob and Jonny first made the otter's acquaintance. As with so many things in this lovely wild world, my father engineered my first meeting.

We were hunting squirrels at the time in the oak and hickory woods south of town. It was a place Dad loved to ramble, and as we slowly made our way over the hills that October morning, he led me down to the creek and followed its meanders to a certain pool upstream. Finding a comfortable place to sit with a tree to lean against, he told me I'd see something special if I kept an eye out. I don't suppose I owned much more patience than any other nine-year-old, but I knew anything Dad promised as something special would be just that, so I sat still and waited.

And soon a long, sinuous shape drifted down into the pool, a flattish head and an undulating back sculled along by a heavy tail. I knew beavers and muskrats and this was neither. In the center of the pool it arched and dived. I looked at Dad.

"Otter," he said quietly.

I looked back at the pool. In a minute or so a head broke through the surface—pugged and sleek and small-eared, mustached with whiskers around its nose. It swiveled from side to side and ducked under again. Dad pointed silently toward the far bank. Presently the head reappeared, this time followed by a thick neck and long body as the animal climbed the bank on short legs, trailing what seemed an impossibly long, pointed tail. At the top it stopped, looked around, shook off a gasconade of spray, and plunged straight down the bank. I knew the move; it was exactly how my friends and I belly-flopped our sleds on snowy slopes.

It hit the water with a splash, disappeared, surfaced again, and once more clambered up. By then I realized there was a definite groove running down the smooth clay bank, straight and slick as a playground slide. And down again it went, forelegs tucked back and hind legs trailing, skidding on its chest and belly, nose pointed straight for the water below. Over and over it climbed and dove, and I knew exactly what was going on. Every kid knows play when he sees it.

Afterwards I visited that watching-spot for almost ten years, until I went away to college. Usually I saw nothing at all except birds and creek chubs fiddling in the shallows. But every so often the otters would be there, too, swimming and sliding, sometimes two of them and on one occasion four, which I knew by then was a female and her litter of nearly grown pups. It was always a lovely place to be, but when the otters were around it was purely magic. I've watched them since in Louisiana and

Minnesota, Colorado and Alaska, and the magic has been there, too, every time.

As the years passed, coupling everything I could read with what I sometimes saw, I learned quite a lot about these engaging animals—learned, for instance, that otters belong to the family Mustelidae, kin to weasels, martens, fishers, mink, skunks, badgers, and the wolverine. I learned, too, that all the various species make up the subfamily Lutrinae, a word derived from the Latin word for otter, which itself came to Modern English through other Northern European languages. It is *otter* in Dutch and German, *odder* in Danish, *utter* in Swedish, *otr* in Icelandic, and it became *otor* in Anglo-Saxon and *oter* in the Middle English that Chaucer spoke and wrote.

In all these languages—and in Russian, Lithuanian, and classical Greek—the sense of the name translates as "dweller in water," and that's as good a description of an otter's life as you're likely to find anywhere. Otters are the only truly amphibious mustelid animals. Not even mink, no mean aquatic performers, are as well equipped for a life in two elements.

Body shape and upholstery are key.

Otters are beautifully built for operating in the dense, cold, resistant world of water. Their bodies are sleek and streamlined. Virtually nothing projects outward far enough to create drag; ears are tiny, legs short. The tail is very thick at the base, tapers to a point, and in the river otter, is flat on the bottom surface. The tails of some other species are flattened differently, but in any case an otter tail is a powerful rudder. All four feet are fully webbed and, with five toes apiece, provide maximum purchase for swimming.

Like many other animals adapted to cold, wet environments, otters wear two layers of fur—dense, oily underfur next to the skin and a longer, coarser outer coat of guard hairs that may be straight or slightly curly. The undercoat is the key to staying warm and dry, as the fine hairs trap minute air bubbles that keep body heat in and water out; otter fur is among the densest of all mammal pelage. A sea otter's undercoat, as I mentioned in the last chapter, comprises as many as 600,000 hairs per square inch, which is about twice as dense as seal fur.

Add to all this equipment specialized valves in its nose and ears that close when an otter is underwater, and you have a mammal better equipped for life in the water than any but those whose legs have evolved into flippers and fins. None of our other amphibious mammals even come close, certainly not the muskrat nor even the beaver—although some observers have found the temptation to compare them irresistible. In *The Gentleman's Recreation*, published in England in 1674, Nicholas Cox says, "It is supposed by some that the Otter is of the kind of Beavers...the outward form of the parts beareth a similitude of that Beast. Some say were his tail cut off, he were in all parts like a Beaver..."

Cox wasn't known as a particularly good naturalist even in his own time (and apparently had a bad habit of listening to the wrong people besides), but if he'd taken the trouble to actually look at both animals side by side he would have noticed that some parts beareth no similitude at all. Only a beaver's hind feet are webbed, for one thing, but more significantly their teeth tell widely different stories. Beavers wear the chisel-like incisors and flat molars typical of gnawing, vegetarian rodents. Look inside an otter's mouth and you'll find the prominent canines and flesh-shearing carnassials of a carnivore.

Although not averse to a bit of carrion now and then and some vegetable matter during the summer, otters capture and kill most of their food. They relish fish and crawfish more than anything else but also dine on everything from frogs, snakes, salamanders, mussels, and muskrats, to snails, earthworms, and the larvae of aquatic insects.

Alaskan sea otters feed mainly on sea urchins and blue mussels. Those in southern waters are fond of abalone, and in particular are the only mammals apart from the primates known to use tools—all the more remarkable because they have developed two distinct tool-using methods. In one they use a rock held between their forepaws to knock abalone from their moorings; in the other they bring flat stones six or seven inches in diameter up from the ocean floor, place these on their chests as they float on their backs, and use them as anvils on which to hammer open thick-shelled mollusks. Both methods appear to be something sea otters learn by example or experience, since the behavior is most often seen in older animals.

For a footed animal to outswim a fish requires both remarkable agility and some keen senses. Hearing seems to be the least developed. An otter on land relies more upon its nose than its eyes. In the water, with ear canals and nasal ducts closed, it hunts to some extent by sight, but its sense of touch is the most useful of all, centered in the whiskers around its snout and in similar tufts of hair on its elbows. These tactile hairs, called vibrissae, are exquisitely sensitive to variations in water turbulence, reading the location and movements of prey with fine precision. This is especially important to sea otters, who might dive to 300 feet in search of food, and it's why sea otters wear whiskbroom mustaches that always remind me of someone's Swedish uncle.

Otters show some eerily humanlike behaviors, such as the sea otter's habit of shading its eyes with its front paws. River otters often have a look to their faces and a general demeanor that I can only describe as mischievous intelligence. I'm not sure how smart they really are, though I have a notion they're brighter than most other mustelids. But there's no question about the mischief. I don't know of any animal more playful, not even the primates.

River otters amuse themselves endlessly with stones and mussel shells, batting them around in the shallows like kittens, tossing them into deeper water and diving to retrieve them, over and over.

Sliding is classic otter sport, and they seem as happy tobogganing on snowbanks as on steep river and stream banks made slick by the water on their fur. Otters traveling over level snow or ice often alternate their loping gait with fifteen- or twenty-foot glides on their bellies. Although it's an efficient way of covering ground while conserving energy, my strongest impression is that they do it mainly because it's fun. This is the only explanation for their downhill sliding that makes any real sense—although the pastime may serve some secondary function in socialization, because otters frequently get up sliding parties that include several individuals.

Their gregariousness sets them apart from most other mustelids, certainly from the largely solitary weasels and mink that are their nearest kin. In most places population densities are so low that groups, or to use the proper collective, pods, of otters typically are family units. Ironically for such amiable creatures, however, the process of creating otter families looks more like a barroom fight than bedroom dalliance. There's a more detailed account of this in the next chapter, titled "Creatures of the Clan"; here, suffice it to say that mustelid physiology is such that vigorous, prolonged copulation is necessary to ensure conception.

Like some other mustelids, river otters exhibit the phenomenon of delayed implantation. Their fertilized eggs migrate to the uterus, develop from a single cell to the spherelike blastocyst stage, and then remain free-floating with no further development for a period of time. Gestation among river otters therefore is highly variable, from 288 to 388 days, so the pups may be born at any time from January to May. Litter size varies, too, with as few as one pup or as many as five. Two to four is most common. They're born blind and toothless, though well-covered in dark-brown fur.

Sea otters also gestate their young for quite some time, as much as eight or nine months, but delayed implantation apparently does not occur in this species; the single pup arrives with its eyes open and bearing a full set of teeth, which suggests continuous development from conception to birth.

Young river otters grow quickly, but even so, they don't leave the den—which is usually an old beaver lodge, muskrat house, or woodchuck burrow—until they're about eleven weeks old. The female has been their sole caretaker to this point, but once they emerge their father lends a hand in their education. Surprisingly this includes swimming lessons, beginning at about fourteen weeks. Before that the parents sometimes carry the pups on their backs as they swim, but when the time comes to begin their amphibious life the young ones have to be coaxed or bundled bodily into the water.

The adults likely will mate again as soon as the pups are weaned, but the family remains together until the following spring. Sea otter families, which amount to a female and her young, stay together somewhat longer and often include pups of different ages.

As pup mortality is rather low and adults have few natural predators—killer whales and Steller sea lions for sea otters and none in particular for river otters—you might wonder why they're so little in evidence. The fact is, otter populations never have been especially dense—high in presettlement days, but never dense. Otters require a lot of living space; individuals may roam as much as a hundred miles of river and stream, and while they don't mind company, too many in one place puts a strain on the food supply.

Their principal predator historically is man. Otter fur has always been the most desirable of all, more even than sable and mink, and the fur trade accounted for the loss of many thousands of animals. It was a harvest the populations might well have weathered in good order had it not been for the double whammy of habitat loss and degradation as the continent has grown ever more manipulated and damaged by human presence. Sea otters, never particularly wary of man, were trapped and killed by the thousands literally from the time the first European settlements were established on the West Coast. They've been protected by law since 1911, but no laws prevented their habitat from being appropriated for human use. And more recently such environmental disasters as oil spills and other water pollution have kept the remaining populations from expanding.

River otters also have suffered grievously at the hand of man. Hudson's Bay Company records show an average 11,000 otter pelts turned in annually from 1821 to 1908. HBC, known to nineteenth-century wags as Here Before Christ, operated only in Canada, of course, but the story in the United States is little different.

I suppose Missouri is as representative as any state. Reports from the last century describe otters as "abundant," "numerous," and "plentiful," none of which is any good for fixing a notion of real populations—except the records also prove that river otters lived in every watershed in the state, and that's no small piece of country. By the mid-1930s, however, a professional survey suggested the state's entire population to be no more than seventy. Blame widespread desiccation of wetlands, blame overharvest, blame what you will, there were virtually no otters here sixty years ago. The situation elsewhere was much the same. Louisiana with its vast wetlands was about the only place in the country where the animals hung on in appreciable numbers, and those populations came to be a wellspring.

By the late 1970s wildlife biologists had plenty of evidence to show that animals can successfully be reintroduced where optimal habitat exists. White-tailed deer, elk, wild turkeys, ruffed grouse, and black bear were cases in point. I don't know if Missouri was the first state to attempt otter restoration, but I do know it worked. Since March 1982, when twenty Louisiana otters were released into carefully chosen habitat, the program has shown steady if gradual success.

◆

Now more than 300 otters have been transplanted to about fifteen different areas of the state, and each release has established a breeding population.

As it happened, the original release site was in a river basin that was also one of my favorite places to hunt quail in the 1980s, and I made a point of looking for otters every time I went there. I wasn't really expecting to see any and didn't for quite some time, but I found enough tracks and droppings to show they were there.

And then one afternoon Tober and I were walking along the river looking for a covey of bobwhites that always foiled us by flushing to the far side at a point where there was no way to cross without swimming, and I found something that wiped away a lot of years in an instant.

It was an otter slide, just like the one Dad showed me long before. I stood and watched it for a long time, imagining a sleek animal plunging headlong down the bank to hit the water in a zestful splash, seeing it surface and look around in its bright, curious, charming way; stood there until Tober ran up wearing the half-exasperated look she uses to remind me that I've lost track of whatever program we're on.

I called Dad that evening to tell him I'd found something really special.

◆

# 23

## Creatures of the Clan

My grandfather Scott owned two characteristics that quite a few Iowa farmers of his generation seemed to share: he was a deadly shot with a .22 rifle and not the least bit shy about using it on anything he considered a threat to his livestock or crops. But because he also harbored a great respect for wild things and a fine sense of their beauty, he saved the skins for me, each one carefully fleshed, stretched, and preserved with salt—fox, muskrat, coyote, rabbits guilty of depredations in the vegetable garden, even the tatty pelt of a Maltese tomcat who showed up and began sowing discord among Granddad's herd of barn cats.

Thus I had a boyhood collection of animal skins second to none. But the jewel of them all, the one I laboriously stitched to a piece of green felt, was the skin of a weasel Granddad happened to see emerging from the hen house at the wrong time. The wrong time for the weasel, I mean.

It was a fascinating little animal, scarcely a foot long, no more than four inches wide slit from vent to chin, a summer pelt white-bellied and brown on top. It was years before I saw another—although I heard plenty of tales about their bloodthirsty nature—and even more years before I learned how such creatures really behave and who their relatives are and where they fit in the scheme of it all.

Mammalogists call them mustelids, members of the family Mustelidae. Together they're a sizeable clan—almost seventy species in twenty-five genera—and in one form or another they live in every part of the world except Australia, Antarctica, Madagascar, and most oceanic islands.

We know them by a multitude of names—weasels, ferrets, mink, martens, fishers, badgers, wolverines, skunks, and otters. Elsewhere the same animals or near relatives are called stoat, polecat, *tayra*, *grison*, *comadreja*, *gluton*, *veelvaraat*, and other such names. They range in size from the least weasel, which, including tail, seldom measures more than about nine inches or weighs more than two ounces, to the largest otters, which may be nearly five feet long and tip the scale at almost eighty pounds. Most have long, slender bodies, but a few are short and stocky.

With only a few exceptions all of them share certain physical characteristics. They all have five toes with prominent claws on each foot, two molars on each side of the lower jaw but only one on either side of the upper. They also are equipped with a pair of anal scent glands they use in maintaining the boundaries of a solitary lifestyle and for defense. Skunks are the most famous for this and possess the most sophisticated delivery system, but certain weasels secrete a musk so nauseating that it makes a skunk smell like gardenias.

Delayed implantation is part of most mustelid reproductive cycles. In this phenomenon, which is also called embryonic diapause, each fertilized egg divides normally until it reaches the form of a hollow, fluid-filled sphere called a blastula; at this stage it ceases developing and simply floats freely in the uterus. The period of dormancy may last several months until some stimulus prompts the blastulae to fasten themselves to the uterine wall and resume developing. No one knows exactly why, although the process must surely be a mechanism by which mating can take place when adult animals are at the peak of physical condition and by which birth can occur when environmental conditions offer young the best chance of survival.

Although the actual gestation period ranges from thirty-nine to sixty-five days, as much as a full year may elapse between conception and birth. Not surprisingly, most species bear only a single litter each year, and a rather small one at that.

Delayed implantation isn't the only reproductive quirk that sets mustelids apart from most other mammals. Courtship, almost without exception, resembles more the violent dance the French call *apache* than anything we usually think of as pitching woo. The male gives the female a thorough thrashing, shaking and dragging her about by the scruff of the neck while she shrieks and carries on as if she were being disemboweled. Copulation, to which all this is a prelude, may last an hour or more, and breeding pairs go through the whole process many times over the course of a few days.

Such apparent savaging of a mate who's always smaller than he is doesn't mean the male mustelid is an unfeeling bully. The fact is, her reproductive physiology requires such behavior; a female mustelid ovulates only in response to vigorous, prolonged copulation, so the whole thing is designed to ensure a high level of certainty that eggs will be produced and fertilized. To accommodate the drawn-out process, nature thoughtfully provides the male with a baculum, or penis-bone. The size, weight, and shape of this bone is for most species a highly accurate indicator of the animal's age.

The mustelids generally are carnivorous, some of them entirely so. A few others will eat anything from insects to meat to vegetation, and at least one has a particular taste for carrion.

All told about fifteen different species are native to North America; some are unique to this hemisphere. As with mustelids everywhere, ours cover a broad range of habits and habitats. Most are mainly active at night, but some are just as happy to go about their affairs by day. Habitats run the gamut from pine woods, deciduous forests, and grasslands, to streams, rivers, and even the ocean. The majority are terrestrial. A couple are enthusiastic diggers and burrowers. A couple spend much of their time in trees. One is semi-aquatic. River and sea otters are highly aquatic. And skunks are like the fabled two-ton gorilla—they live pretty much wherever they want.

Apart from the striped skunk, none of the North American mustelids is especially plentiful, and a few are in jeopardy of disappearing altogether. Without exception they are unfailingly interesting.

As there are chapters devoted specifically to skunks and otters, I won't rehearse their stories here. But as to the rest...

**Weasel**

In times past taxonomists have recognized as many as thirty-five subspecies of weasel in North America. Current thinking reduces the number to three: the least, short-tailed, and long-tailed weasels—*Mustela nivalis, M. erminea,* and *M. frenata,* respectively.

They are the smallest of the mustelids. The least weasel in fact is the smallest of all carnivores (larger, of course, than shrews, but shrews are not strict meat-eaters), and not even the burliest long-tailed weasel is larger than a gray squirrel. Like terriers, though, they're larger on the inside than the outside. They're all deadly mousers, and the two larger species are easily capable of dispatching prey considerably bigger and heavier than themselves, including cottontail rabbits and domestic chickens.

Hen-house indiscretions are invariably the work of specific individuals, but such antics have given weasels a bad reputation in some quarters—that and their penchant for killing more than they eat. Some is cached away but much goes to waste, so you don't have to look far to find someone who'll describe a weasel as a wanton killer. Common belief has it that they suck their victims' blood, a myth no doubt prompted by their habit of licking blood from the wounds of freshly killed prey. This notion was so widespread years ago that even some professional mammalogists bought it. But weasels are not musteline vampires.

Some species' seasonal change of color probably has earned them even more grief than their dietary habits. In northern climes weasels turn white in winter—except for the tips of their tails, which are black year-round (those of least weasels have only a few black hairs at any time). In the white phase they're called ermines and have for centuries been highly prized in the fur trade.

Such radical shifts in coloration require two molts every year, and the transformation differs with the season. In autumn a weasel molts from the belly upward and for a while wears a mottled coat of brown and white; the spring molt starts at the top of the back with a distinct line of brown fur that slowly works its way downward.

Delayed implantation occurs in both the long- and short-tailed species, but the least weasel is able to breed at any time of year and typically bears two or more litters annually.

**Black-footed Ferret**

*Mustela nigripes* is a close relative of the weasel and in its genus second only in size to the mink. A ferret is as large as some fox squirrels. Unlike the weasel, it doesn't change color in winter.

The black-footed ferret is unique to this continent. Some mammalogists believe its ancestors were steppe polecats that migrated from Asia via the Bering land bridge during the Pleistocene, and they may be right. Our ferret and the Asian polecat are both grassland animals fond of dining on a variety of burrowing rodents, able by virtue of their lithe, slender bodies to travel labyrinthine burrows in quest of prey.

The prairie dog is by far the most common prey animal on the Great Plains, or at least it was when the black-footed ferret was evolving its lifestyle. But the ferret's dependence upon a single species of prey has led the predator to the brink of extinction.

Prairie dogs are highly gregarious, living in vast towns that may cover hundreds of acres. Unfortunately the numerous

burrows are leg-breakers for cattle and horses, so for many years ranchers went on sprees of prairie-dog poisoning to make the West safe for beef, always with wholehearted approval (and frequently direct assistance) from the federal government. As the prairie dog's population dwindled, black-footed ferrets lost both a source of livelihood and the appropriated burrows they lived in.

Although the prairie dogs rebounded vigorously when the poisoning stopped, ferrets did not. They were nearly extinct 50 years ago and unfortunately are no better off now. It's a pity, because we essentially lost an interesting little animal before anyone thought to learn much about its habits or its life.

**Mink**

As the steppe polecat made its way along the Bering isthmus toward North America sometime between 10,000 and 2 million years ago, it may well have passed a relative heading the other direction. All evidence suggests that mink originated on this continent and that some individuals migrated westward to the Old

World during the Pleistocene; there it became *Mustela lutreola,* the European mink. The North American species is *M. vison*—which, as if to make sure no one misses the point, translates as "weasel that is a kind of weasel."

The mink represents something of a bridge between the most diverse habitats that mustelids occupy. Like the weasel it lives on land, hunting mice, rats, rabbits, and other small animals, but it's almost equally at home in water, swimming like an otter after fish, frogs, crawfish, turtles, and the occasional duck.

It combines some diverse physical characteristics as well. The mink's long, slender body is essentially weasel-like though proportionately stockier, and the five toes on each foot are partially webbed—not as much as the otter's, but better adapted for swimming than the weasel's.

Mink seldom stray far from water, living in bankside dens, in hollow trees, under logs or stumps, even at times in muskrat houses whose former occupants ended up as a good meal. They're most active at night, often stirring about at dusk and dawn. Seeing a mink in broad daylight is a rare treat. Drifting some wild river in a canoe is usually the best opportunity, but even then a glimpse is often all you get; they're quick, quiet, and remarkably agile, slipping in and out and through the water with scarcely a ripple.

**Marten**

At first glance a marten looks like the product of a *ménage à trois* involving a mink, a fox, and a housecat. Males measure up to two and a half feet in body length with another eight to ten inches given to tail, and they stand about seven inches tall at the shoulder. As in all mustelids, females are only three-quarters as large and weigh perhaps a third less.

Of seven species worldwide, only *Martes americana* is native to the New World. Here it is sometimes called American or Hudson Bay sable, suggesting its kinship with the Asian species so highly prized in the fur trade. And like one of the European species it's also known as pine marten, which is an apt description for its habitat.

Wherever they live martens are strictly forest dwellers, often in pine woods, and the denser the growth the better they like it. As mink divide their time between land and water, martens divide theirs between ground and treetops. They're extremely nimble climbers, skillful at catching red squirrels, which are their main prey wherever squirrels are abundant. Otherwise they're content to feed on mice, hares, grouse and other birds, insects, fruits, even carrion.

Whether a marten spends most of its time aloft or aground depends on the season. It does more tree climbing during the summer. In winter it's fond of tunneling under snow, no doubt hunting mice that are doing the same thing.

Extremely shy, martens steadfastly refuse to live near much human activity, so there are lots of unanswered questions regarding their lifestyle. They seem to be active day or night and live largely solitary lives. Delayed implantation is well documented in most species, so they rear only one litter each year.

**Fisher**

The fisher, sometimes called blackcat or Virginia polecat, is the largest of all the martens, measuring as much as fifty inches from nose to tail-tip. A male might weigh fifteen pounds.

*Martes pennanti*, which is unique to North America, is in many respects similar to the marten. It likes the same sort of forest habitat and occupies much the same geographic range. Its breeding habits are similar as well.

Fishers do not appear to spend quite as much time in trees as martens, though they're every bit as agile at climbing. The most intriguing difference between the two is diet, for the fisher's favorite prey is the porcupine. Its agility is therefore a good thing, because even though porkies are neither physically nor mentally speedy, they do have some almighty effective defenses, and dispatching one without ending up stuck full of quills requires as much finesse as successfully tap-dancing across a minefield.

Obviously the standard musteline means of killing prey with a bite to the back of the neck wouldn't work, but prickly as they are, porcupines do not grow quills on their faces or bellies. Those are the fisher's points of attack—either sending the porky into shock with repeated bites to the face or, through some tactic of feint and maneuver, bowling it over so its vulnerable belly is exposed. A fisher seems able to kill a porcupine in a tree just as handily as on the ground.

But it's no easy job anywhere, and it can take a half-hour or more before the fisher can begin its meal. It always begins feeding on its victim's belly, eating bones and all over the course of a week or two, eventually leaving nothing but a skin spread quill-side down like a spiny bath mat.

Popular belief hails the fishers' unfailing ability to kill porcupines with never a prick from a single quill. The fact is they often get stuck. Some die of infection as a result, and virtually every fisher pelt shows the scars and flattened quill-tips to prove that life's tough no matter how nimble you are. Some "reports" even describe how fishers kill porcupines in winter by burrowing up through the snow to get a clear shot at the belly, but that's a notion fetched a bit far for me.

**Badger**

Apart from the otters, none of the mustelids appear quite so amiable as badgers. Short-legged, pigeon-toed, and waddle-gaited, wise of countenance and with a catlike perk to their ears, both the American and Old World species of badgers look like easygoing chaps dressed in clothing tailored for some much larger animal.

Such appearances are a bit deceiving. Badgers are reasonably laid-back if they aren't molested, but they can also be grumpy and quick to take exception if they feel put upon. Pick a fight with one and you'll wish you hadn't. They aren't very big—nine or ten inches tall at the shoulder, about twenty-eight inches long, seldom weighing more than twenty pounds—but they're incredibly tough and, as more than a few dogs have discovered, can be a world of trouble in a small package. Their African cousin, the honey badger, is even shorter-fused, known to attack other animals and even humans unprovoked.

In the Northern Hemisphere, badgers are mainly grassland dwellers, finding both their living and their security underground.

Although happy to eat almost anything from eggs to insects, lizards to scorpions, the New World version is a great hunter of small rodents—ground squirrels, prairie dogs, rabbits, and such. Sometimes they dig their prey out of burrows and at other times lie in ambush. Lore—not fully documented but probably at least partly true—has it that coyotes and hawks occasionally follow foraging badgers on the chance of picking off a meal that happens to get away.

Surprise one near its own burrow and a badger is underground in a flash, equally adept at going in backwards or headfirst. Catch one that's wandered any distance away and if you don't crowd it too much it'll still go underground, simply dig itself in, muttering and growling as it works.

A badger's long, heavy, curved front claws are splendid earth-moving tools driven by powerful shoulders. It typically uses its front feet to loosen the dirt and its hind feet to kick it out. In loamy prairie soil a badger intent on escape can dig faster than a man with a shovel.

For all their curmudgeonly ways badgers have a decided charm about them. Years ago a friend reared one from infancy to keep as a pet. He had the appetite of a horse and you had to be careful not to startle him, but Tax was otherwise a delight. (The name came from his scientific moniker, *Taxidea taxus*, which unpoetically enough means "badgerlike badger.") Sitting around one evening I heard a soft, sweet sound that for a moment I couldn't place. Then I realized it was Tax, dozing on the sofa next to me, purring. I've had a soft spot for badgers ever since.

**Wolverine**

Few North American animals are more laden with myth and fancy than *Gulo gulo*. From old trappers' tales come fabulous stories of wolverines leveling log cabins, beating up grizzly bears, and attacking humans with a ferocity little short of diabolic. The hell of it is, most of them hold at least a grain of truth.

Except for sea otters, the wolverine is the largest of all the mustelids. A big male might be nearly four feet long in the body, stand fifteen inches at the shoulder, and weigh almost sixty pounds. His musculature is massive and his strength prodigious—one of the most powerful animals of his size in all the world. Combined with the utterly fearless nature typical of mustelids, it's not hard to see how tall tales might accrue.

Properly motivated—say, by a cache of food inside—a wolverine could easily demolish a flimsily built hut. It could do as much on a whim, because ferocious as it is when cornered, a wolverine left to itself is as playful as an otter, and as the old saying goes, they don't know their own strength. I suspect more than a few reports of *Gulo*'s epic destructiveness originated with some wolverine's fancy to play hockey with a trapper's or camper's belongings.

This is not to suggest that they're cuddly and sweet. Wolverines are reliably known to sometimes appropriate kills from mountain lions and grizzlies by the sheer force of determination. That, too, is typical of the mustelids. In winter, buoyed on feet that are extremely broad in proportion to their body weight, wolverines can easily dispatch snow-mired deer and even elk—and a fifty-pound animal that can take down an elk under any circumstances is a serious customer.

So is one given to raiding wasp nests to feed on the larvae, and this is how a lot of wolverines make their late-summer livings. Otherwise, they like bird eggs in early summer, berries of several kinds in the fall, and carrion whenever they can find it.

Wolverines are creatures of wide-open spaces. A male's home range may be 750,000 acres or more. He's happy to share his turf with a few females, though finding one another at the appropriate time must take some effort. But they do, and delayed implantation is part of the cycle that culminates in the birth of two or three kits in late winter or early spring.

Along with being nasty tempered, wolverines are also widely thought to possess an appetite of such proportion that "glutton" is a common nickname. One early report, written in 1518, says the animal customarily "eats until the stomach is tight as a drum skin then squeezes itself through a narrow passage between two trees. This empties the stomach...and the wolverine can continue to eat until the carcass is completely consumed." Exactly what prompted this tale is anyone's guess, but I do know it isn't true. Wolverines are hearty eaters but no more so than any other carnivore.

Bizarre as it is, the image of a bulimic wolverine isn't really any more outrageous than some things once, or even still, believed about any number of other animals. As often is the case, though, some of the facts are every bit as astonishing as any of the fictions. Certainly this is true of the mustelids, as a family and as individual species. They're proof that nature has the best imagination after all.

◆

# 24

## Little Brother to the Bear

The first European colonists to make landfall in the New World found a wilderness rich in animals. Some, like deer and bear, were familiar enough. Others, like the strange, slow-witted beast the Algonquian Indians called *apasum*, were entirely new. Yet others simply were confusing—and none more so than the roguish-looking, ring-tailed creature the Indians knew as *arocoun*.

*Apasum*, who obligingly fainted when someone picked it up by the tail or scruff and was therefore easily shipped back to the Old World, soon found itself the center of great sensation as Europeans got their first look at a marsupial. *Arocoun* proved less cooperative, eluding capture by shinnying up a tree or simply biting the hand that grabbed it, so the early naturalists made their observations from a respectful distance. The confusion began in their attempts at describing *arocoun* in terms of animals they already knew.

Captain John Smith published the first description in 1612, remarking upon "a beast they call *Aroughcun*, much like a Badger, but useth to live on trees as squirrels doe." By 1743 *arocoun* had been adopted into English as *raccoon*, and John Bickell, who in that year published his *Natural History of North-Carolina*, had this to say: "The *Raccoon* (which I take to be a species of *Monkey)* is of a dark grey Colour, and shape and bigness partly resembles a *Fox*, but has large black Eyes, with great Whiskers like a Cat, the Nose like a *Pig*, and the Feet are form'd like a Hand or those of a *Monkey*...they are easily made tame and familiar like a Dog, yet they are very Apish...."

No one else did much better. Rather than adopt an Indian word, French explorers dubbed the raccoon *chat sauvage*, or "wild cat." But in *Histoire de la Louisiane*, 1758, Antoine du Pratz took exception: "The *chat sauvage* was improperly named...for it is like the cat only in its suppleness, and actually is more similar to the Marmot...Its head is somewhat like the fox's."

A few years earlier Karl von Linne had concluded that the raccoon was some sort of bear and named it *Ursus cauda elongata*, "long-tailed bear." In 1758 he changed his mind and renamed it *Ursus lotor*, "washing bear." Colonial American naturalist Mark Catesby thought it was related to the fox and therefore named it *Vulpi affinis americana*.

About 1780, German naturalist Gottleib Conrad Christian Storr decided that everyone was wrong, that raccoons belonged in a class of their own, and gave their genus the name it still bears—*Procyon*. Storr had the right idea biologically, but his choice of generic name created yet another level of confusion. *Procyon* comes from two Greek words that combine to mean "before the dog," which had led some to infer an ancestral connection between raccoons and dogs. Actually, the Procyon Storr had in mind is a star, and it rises just before Sirius, the Dog Star. So why name a genus of animals after a star? Unfortunately it's too late to ask Herr Storr, but even if we could his answer probably would be "Why not?"

◆

Raccoons, it would seem, simply were not meant to be easily understood.

They do have some close relatives—the lesser panda, the ringtails, cacomistles, coatis, coatimundis, kinkajous, and olingos. At slightly greater biological distance, they're even related to the

giant panda. Except for the two pandas, which are native to eastern Asia, all of these animals belong to the Western Hemisphere. Since European naturalists knew of virtually none of them before the nineteenth century, their dithering about badgers, marmots, foxes, monkeys, pigs, cats, squirrels, bears, and dogs wasn't so farfetched after all.

Besides, raccoons really do show certain traits, either physical or behavioral, that are bearlike and foxlike and doglike and like all the other creatures that science once supposed them to be. Their hind legs are longer than the front ones, which gives them something of a humpbacked, bearlike gait. Also like bears (and humans), raccoons are plantigrade, walking on the soles of their feet instead of on their toes, and though technically classed as carnivores they are omnivorous in practice, able to make a meal of just about anything.

◆

They climb trees like squirrels and cats, have sharp, pointy noses like foxes, develop heavy, loose layers of body fat like badgers and marmots, and have jaws and teeth that look much like a dog's. Most characteristic of all, they are as brightly intelligent, as insatiably curious, and nearly as deft with their front paws as any monkey.

Such things considered, it's small wonder that early observers found *arocoun* a confusing animal, or that they had ample opportunity to be confused, since raccoons lived in great numbers throughout the original thirteen colonies and, for that matter, virtually throughout North America from southern Canada to Panama. One species, the crab-eating raccoon, occurs from Costa Rica south to Peru and Uruguay.

As you might expect of any animal so widespread, raccoons come in a variety of forms, and at times as many as thirty-one species and subspecies have been recognized. For the most part the differences have mainly to do with size and coloration. Raccoons in the Florida Keys might weigh no more than five pounds at the most, and specimens come progressively larger the farther north you find them. From Missouri on up into Canada average weight is about fifteen to twenty pounds among males, a bit less for females. The current heavyweight record-holder came from Wisconsin and weighed in at just over sixty-two pounds.

In temperate climates raccoons are active year-round. Those in the north live more cyclical lives, adjusting their activities according to the weather and changing their diets according to what foods are most plentiful. Although they tend to go for high-protein victuals during the springtime mating season, and highly caloric grub in the fall, raccoons are supremely democratic trenchermen whose favorite meal tends to be whatever they're eating at the moment. It could be fish, frogs, crawfish, mice, chickens, eggs, nestling birds, newborn rabbits, freshwater mussels, clams, oysters, wild or domestic fruits and berries of all kinds, corn, nuts, acorns, crabs, turtle eggs, garden vegetables, insects, worms, or garbage. If it's edible at all, it's coon food.

◆

In northern latitudes raccoons go on an eating binge through the autumn, and it's serious business. If they don't build up a thick layer of fat they won't survive the winter, because they do not hibernate. They den up and sleep during the coldest weather, but their body temperature, heart rate, and metabolism all remain at much higher levels than those of true hibernators. Adults may lose fully half their body weight by spring.

Whatever explanation this offers for the fact that a raccoon is a walking appetite is only half the story, for even warm-climate coons are perpetually on the prowl for something to eat. Given their catholic taste and ingenuous curiosity, scarcely any animal alive is more opportunistic or more adaptable at living close to man.

Which they do in surprising numbers, thriving in small towns, in suburban neighborhoods, even in great, sprawling cities. Washington, D.C., which covers an area of sixty-eight square miles, is home to an estimated eight thousand raccoons, and the populations in other eastern seaboard cities may be higher still. The animal's quiet nature and largely nocturnal lifestyle make accurate counting impossible, but there's no question that the resourceful raccoon can make a good living in our cityscapes, finding shelter in parks and buildings and dining happily on the astonishing quantities of garbage that all cities create.

Since they seldom come in contact with people, no one gave them much thought until the late 1970s and early '80s, when a wave of rabies swept through the eastern United States. What began in 1977 with one rabid raccoon in West Virginia was by 1983 an epizootic—1,608 confirmed cases among urban raccoons in Virginia, Pennsylvania, Maryland, and the District of Columbia. Since then the numbers have declined and risen in cycles typical of the disease.

Rabid raccoons pose little direct threat to humans—except to those who try keeping them as pets or who attempt to handle any wild animal; the one is illegal, the other merely stupid. The indirect hazard, however, is more serious, since unvaccinated dogs and cats that make contact with any rabid animal can act as intermediaries in passing the disease along to people. Researchers at several institutions have worked for years on synthetic rabies vaccines that can be administered to wildlife through some sort of bait, but none of these efforts have progressed much beyond the laboratory stages.

Even though they're perfectly content in coming to us, raccoon watching is a far pleasanter pastime when it's done on their turf. I live in a state that has some of the wildest, most beautiful rivers on earth, and there's no better place to meet raccoons than a gravel bar on a moonlit night.

Except in the dead of winter, a canoe trip down any south-Missouri stream would be incomplete without the raccoons' nightly parade out of the Ozark hills and into camp, rummaging through my gear, chittering and chirping to one another as if discussing the relative merits of whatever they find or the logistics of how to pick the chopbox latch. Anything they can't eat is likely to be hauled off as a plaything if it's small enough to carry, and I've been thus robbed of silverware, can openers, matches, tinfoil, candles, and a good English briar pipe. I woke up one night to find a furry little arm reaching through a rip in the mosquito net, trying to fish my wristwatch out of the possibles-pocket sewn to the wall of my tent. Perhaps it's the bandit's mask they wear or the saucy, unblinking stare from those shiny, jet-black eyes, but a coon caught in some miscreant act looks at you as if it's sowing mischief just for the fun of it.

Before I wised up and built a coon-proof box for provisions, the Current River population prospered on my eggs, bacon, apples, cheese, salami, and bread, trundling their snacks one by one down to the river's edge, there to perform the most widely known of all raccoon rituals.

Karl von Linne chose the epithet *lotor,* "washer," for the raccoon's scientific name after watching captive animals dunk their food in water. Later observers were so intrigued by the trait—and so strongly influenced by von Linne's belief in the kinship of raccoon and bear—that the animal is known virtually throughout Europe as the washing bear: *tvättbjorn* in Sweden, *waschbär* in Germany, *oso lavador* in Spain. In Russia, where raccoons were established in the 1930s, they're called *enot-poloskun,* the latter part of which means "rinser." The French, being French, took a slightly different view and called the raccoon *raton laveur,* "washing rat."

While a raccoon is neither bear nor rat, he is decidedly a washer, even though we don't know exactly why. Early naturalists, products of an age inclined to interpret animal behavior in human terms, took the raccoon's food-washing literally as a sign of cleanliness and therefore a good example. Scientists early in this century, determined to be scientific, proposed a more rational view—that raccoons dunk their food as compensation for underactive salivary glands. This theory went out the window in the 1960s when researchers determined that *Procyon lotor* manufactures all the saliva it needs without any artificial help.

So if raccoons are neither spitless nor especially tidy, why do they immerse their food?

Actually they immerse all sorts of things, not just food, and dunking is not a prerequisite for eating; if it were we wouldn't find desert populations surviving nicely on juniper and manzanita berries and keeping themselves hydrated by licking dew. Even raccoons that live where water is plentiful eat the majority of their food right where they find it, and they find most of it a long way from water.

But if water is handy a raccoon can't seem to resist dunking any object it finds. All the various forms of "washer" aside, the original Algonquian name describes the animal best, for *arocoun* means "scratcher" or "feeler." While its sense of hearing is good enough, neither sight nor smell are particularly acute, and a raccoon's main contact with the world is through touch. Watch any wide-awake coon for a while and you'll notice that its front paws are constantly busy, picking up small objects, edible or otherwise, rolling them around, dropping them in favor of something else, endlessly dabbling and feeling. Look closely at the paws and you'll find five slender, fingerlike toes with smooth, hairless pads underneath; the soles also are bare and padded.

The only significant difference between a raccoon's paws and a primate's hands is that coons have no thumbs—which, if you'll pardon the pun, doesn't seem to be much of a handicap. Backed by a remarkably high level of problem-solving intelligence those little paws are clever as magic, able to deal succinctly with a whole array of latches, catches, windows, lids, and locks.

Water no doubt heightens the animal's already sensitive touch, and what looks like a fetish for washing things probably is instead an attempt to extract every bit of tactile information. A raccoon seldom looks at whatever it's fiddling with, in or out of water; instead it stares fixedly elsewhere, as if bringing all its concentration to bear upon interpreting the messages of touch.

Given their inquisitiveness and penchant for feeling things, raccoons are as easily lured with some shiny object as with bait, a fact that fur trappers have long used to advantage. To Indians and colonists alike *arocoun* was a source of food and fur, and he remains so today. The flesh is dark, a bit on the greasy side, and from an older animal somewhat tough and strong-tasting. But a young coon properly prepared—well trimmed of fat and roasted—is delicious.

The raccoon's greatest economic clout, however, has always come from his pelt, which comprises a dense, dark-colored coat of underfur and longer, coarser guard hairs, each one banded in black and tan. Such handsome dress serves the animal with both insulation and camouflage and has served man in a multitude of uses. It was headgear in the days of backwoodsmen and long hunters—and again in the mid-1950s, when several million Americans went half mad on Davy Crockett.

At the turn of the twentieth century raccoons supplied the raw material for all sorts of stoles, muffs, boas, capes, carriage rugs, and busbies. Raccoon coats were *de rigueur* among Ivy League college boys of the 1920s, and the fur has several times since been in and out of fashion as coat collars and trim.

Unlike mink, raccoons cannot be profitably pen-reared for the fur trade, but not because they take poorly to captivity. Quite the contrary. A caged raccoon develops a gargantuan appetite, and by the time it's grown large enough to produce a salable pelt it has eaten several times the value of its own skin. Some of the roars in the Roaring Twenties came from would-be coon ranchers who learned this the hard way.

Consequently, nearly all raccoon fur converted to human use has come from the wild. At certain times and in certain areas harvest has taxed populations severely but never permanently. Like everything else associated with the world of fashion, the fur market follows a goofy, seesaw course; prices are up one year, down the next and, especially of late, are often depressed for years at a time. When that happens, the harvest naturally declines. Under modern systems of game management and harvest regulation even a fur-market boom isn't likely to have any serious impact.

Raccoons have supplied other uses besides fur and food, especially in earlier days when we were a considerably more frugal, nature-oriented society. On the eastern frontier coon lard preserved and waterproofed leather boots almost as well as bear grease. Raccoon fat could lubricate a wagonwheel or be rendered into lamp oil. From 1784 to 1788 coonskins passed as currency in the short-lived state of Franklin, an area between the Cumberland and Blue Ridge mountains in what now is east Tennessee. Five hundred coon pelts represented the annual salary of the secretary to His Excellency the Governor, and the clerk of the House of Commons earned two hundred per year. (His Excellency himself drew the princely sum of a thousand deerskins per annum.)

One curious little item remained useful long after raccoons ceased to serve either as lubrication or legal tender. Like many other animals—all of the rodents and most of the carnivores—a male raccoon's skeletal structure includes a baculum, a penis-bone.

Some female rodents and carnivores possess the counterpart, a tiny bone called the *os clitoridis.*

Mark Catesby was the first to discuss bacula in print, and naturalists for many years assumed that raccoons were the only animals to have them. (This along with the opossum's bifurcate, or forked, penis no doubt caused some puzzlement among early mammalogists on both sides of the Atlantic.) Now mammalogists have learned that the size, shape, and weight of bacula are highly reliable indicators of age in raccoons, mink, and other species.

An adult raccoon's baculum is a hook-shaped, tapered bone at least three and one-quarter inches long, and it has served a multitude of homely uses. In colonial times and for many years after, frontiersmen and farmers were fond of using them to ream ashes from their pipes. Townsmen wore them as watch-chain ornaments. Hunters fastened a steel needle in one end and used it to clean the flash-channels of their flint- and caplock rifles. At one time every country woman in the South kept a baculum in her sewing basket, because the natural notch in the hooked end makes it an ideal implement for picking stitches out of cloth.

Folklore and superstition inevitably adopted them, too, and besides their utilitarian roles bacula have been used as charms, *mojos,* and *conqueroos.* They also have been an endless source of pranks and practical jokes. My father had an particularly large, highly polished specimen that he sometimes carried in a shirt pocket. Standing at the Elks Club bar, he'd fiddle with it absent-mindedly until someone asked what it was—to which he'd reply that it was a dog-whistle and hand it over with an invitation to give it a toot. The victims almost invariably blew themselves cross-eyed trying to produce a sound and some refused to believe him when he told them what it really was.

There's a certain symmetry in all that, because the animal itself has long been a folklore figure of mischief and fun. As *arocoun, shoui, wayatcha, eseeban, atuki, asipun,* or any other of the names by which the Indians knew him, he appears in their legends and tales as the Trickster, the rascal, the perennial hoaxer, his life a picaresque tale of jokes and comic deceit. As raccoon he plays a similar part in later stories from tall tales to Mark Twain. In these he is less anthropomorphic, but whatever is outlandish, beguiling, or mischievous in American folklore is likely to have a raccoon mixed up in it somewhere.

Although the western Indians, relying upon appearance, identified the raccoon with a word that translates as "younger brother to the bear," almost everyone who observes his behavior sees him as a paragon of cleverness. The notion that raccoons fish for crabs by dangling their tails in the water is so old that it was reported (as fact) by John Lawson in 1715—but you wouldn't have to look far today to find someone who would tell you in all seriousness that this is precisely how coons catch crawfish. It isn't true in fact, but it's accurate enough in spirit. Any animal that can brook nearly four hundred years of human interference and still flourish as the raccoon does clearly has more going for it than just a pretty face.

# 25

## The Prince of Briars

Once the great Creator god made the buffalo and the deer, he made grass and festooned the trees with tender green leaves to give them nourishment. When the first eagle flew, he set the waters teeming with fish, hung trees and bushes with fruits to feed the *apasum* and *arocoun,* made mice for the hawk and fox.

Then he retired to a promontory to survey the wonderful tapestry of life he had created, where presently a delegation of animals approached—coyote and hawk, bobcat and owl, fox and crow, mink, weasel, snake, and others.

"Do not think us ungrateful," the crow said, "but our livelihood is incomplete. You have given us the need to eat flesh and yet provided little upon which to prosper.

"There is food in abundant variety for those whose lives depend upon greenery; you have made deer and moose for the cougar and wolf. For us there are mice but little else, and mice are very small."

The Creator god listened in silence, studying the animals before him.

"You speak the truth," he said at last. He breathed into his cupped hands, capturing the essence of life, molded it for a moment, and brought forth a rabbit.

"Behold," the Creator god said. "I give you bounty."

You won't find this story in any collection of myths from any culture, because I just made it up.

But it's valid nonetheless, or would be if biology were given to mythmaking as a way of describing the complex interrelationships in what often is referred to as the food chain, which originates in chemical reaction and the activity of microorganisms and becomes visible nature with the presence of plant life. This is converted to animal matter by a vast variety of creatures—bird, insect, mammal, even a few reptiles. These in turn nourish the flesh-eaters and insectivores. Among mammals, mice and other rodents of similar size form the crucial link between plants and carnivores, but in some parts of the world the next niche up in size belongs almost solely to the rabbit.

Certainly this is true in North America where rabbits are a major source of food for virtually every predator larger than a least weasel and smaller than a wolf—including the birds of prey and some snakes. If the predator's rôle is to harvest a portion of other animal populations, the rabbit's lot is one of contribution.

Among other things, the rabbits contribute a measure of confusion. They all look essentially alike, but some are rabbits and others hares. As a group they once were considered members of the rodent family, largely because of similar dentition—chisel-like incisors separated from the cheek teeth by a wide space called the diastema. Unlike other rodents, however, rabbits and hares grow a second pair of upper incisors directly behind the first and now are classified in an order of their own—Lagomorpha, which means "hare-shaped." They belong to the family Leporidae and, in North America, to the genera *Lepus* and *Sylvilagus,* three names that translate, respectively, as "hare," "hare," and "wood hare." To make the tangle complete, about half our native hares are commonly called rabbits.

Biologically the differences are few but significant. Rabbits are born blind, nearly naked, and helpless; hares arrive with eyes open, well covered in fur, and able to move about shortly after birth.

Both forms are well represented among North American fauna, with seven species of hares and eight of rabbits. Native hares, which all belong to the genus *Lepus*, include the varying or snowshoe hare of Canada and the northern United States, the tundra hare of the high Arctic, and four species commonly called jack rabbits—white-tailed, black-tailed, antelope, and Gaillard's. The seventh species is *L. europeus*, the European common hare, imported to the New World between 1888 and 1911; it now occupies considerable range along the eastern seaboard and west around the Great Lakes.

Like the hares, North American rabbits also belong to a single genus, *Sylvilagus*, and the species are distinguished more by size, geographic location, habitat, and minor color variations than by striking physical differences. Thus we have the pygmy rabbit, brush rabbit, marsh rabbit, swamp rabbit, New England cottontail, mountain cottontail, desert cottontail, and most widespread and familiar of all, the eastern cottontail.

Rabbit or hare, one species or another, they're all variations on a single theme: short, powder-puff tail; long ears; long hind legs with big, fur-soled feet; a divided upper lip beneath a blunt, velvety-furred nose; and relatively large eyes set well to the sides of the skull. The females are larger than the males, and the species range in size from the white-tailed jack at as much as ten pounds to the pygmy rabbit at a pound or less. The European hare can grow as large as twelve pounds, and while hunting tinamou in Argentina I've flushed some that look as if they'd go a pound or two heavier still. To an Iowa farm boy accustomed to cottontails, a rabbit the size of a dachshund is not something soon forgotten.

◆

Nor for that matter is a boyhood seamed with rabbit trails, so I trust you'll understand if I think of the eastern cottontail as Everyrabbit and focus this piece mainly on the gentle little chap who sits steadfastly in my mind as the prince of briars.

For an animal whose main function seems to lie in the business of converting grass and clover into food for coyotes and bobcats and red-tailed hawks, the cottontail accepts his role with an extraordinary measure of grace. And it's not that he makes himself easy fare. He has a sharp sense of smell (which is why his nose is ever a-twitch), excellent hearing, a field of vision that exceeds 360 degrees, nerve enough to sit absolutely still while fully exposed, and when he needs it, the ability to quit the scene at high speed.

The pygmy rabbit tends to scurry rather than jump, but all the rest are great leapers. A cottontail can cover five feet at a bound, and a jack rabbit at full stretch at least twice that much. Few predators can match pace with the big hares—certainly none of the bird dogs I've watched test their legs against jack rabbits, who seem to view the whole business as more lark than struggle.

And a six- or seven-pound body traveling at high speed can generate some impressive ballistics, as I discovered in the 1950s when a jack matched up with the headlight of my uncle's Lincoln while both were doing roughly ninety miles an hour across the New Mexico desert in opposite directions. The hare paid dearly for his miscalculation, of course, but so did the headlight, which ended up resting against the windshield frame along with the highly compressed remains of what had formerly been a perfectly good fender. Meetings between rabbits and automobiles aren't usually so spectacular; most often the result is simply one more meal for the scavengers who have a taste for leporine flesh.

Beset by peril from so many quarters, rabbits naturally have evolved means of ensuring that their species survives. Their sensory capacities are acute, as I mentioned earlier, while their behavior and choice of habitat further minimize risk. Cottontails are adaptable creatures, able to get along in almost any environment from wilderness to suburbia, so long as it offers some media for concealment and escape. In the wild they prefer open brushland or the bushy zone of woodland edge, fencerows, tall grass, rank weed patches, brushpiles—whatever provides a protective screen from predators both landbound and airborne.

Like many prey animals they are most active at night and early morning and spend their days tucked away in a resting-place called a "form"—which is simply a small depression scratched out in a clump of grass or under a bush, where a rabbit can lie facing out. Cottontails living in a given area use different forms as the fancy strikes, but a few dominant individuals seem to stake out certain forms as their own. Rabbits often follow the same routes in moving among the forms and on feeding excursions through their few acres of home range, making trails and runs that can be quite easy to spot.

Rabbits and hares alike go about their lives quietly. Cottontails occasionally thump the ground with a hind foot as a warning signal, and females utter a little call to gather their young, but otherwise they conduct business in silence unless frightened or scuffling among themselves in some dispute. When captured, however, both rabbits and hares let go a loud, shrill, whistly scream that's about equally heart-tugging and hair-raising. The sound is so distinctive that a good imitation will bring coyotes and sometimes even bobcats approaching on the run.

A few species of rabbits show some seasonal color variation but for the most part remain the same year-round. Some northern hares, however, grow white pelage for winter as camouflage against the snow. The varying hare, arctic hare, and white-tailed jack rabbit change color thus.

For all the individuals' defenses the most important factor in the species' survival is an ability to breed like...well, like rabbits. In warm climates the breeding season can be virtually year-round, and even here in Missouri the cottontails usually begin mating in mid-February, as do jack rabbits as far north as North Dakota.

Nearly all the Leporidae perform some sort of courtship ritual. For cottontails it's an interesting, highly stylized affair that begins with the male approaching the female from behind. She spins to face him, he makes a rush at her, and she jumps over him in a game of reverse leapfrog. They repeat this several times, occasionally switching roles, until some signal triggers copulation.

Gestation time varies according to species—an average forty-three days for jack rabbits, thirty-seven days for varying hares and swamp rabbits, and twenty-seven days for cottontails. As her time approaches, a female cottontail finds a spot she likes and digs a nest, roughly five by seven inches and three or four inches deep. Since she typically digs from only one end, the opposite end usually is slightly undercut. She lines the cavity first with a layer of grass, then with a layer of fur she plucks from her belly, and finally tops it off with a covering of grass. Sometimes the sites are wisely chosen and sometimes not, for cottontails often dig nests in the open on lawns, parks, and golf courses.

Litters also vary, but four to six is a typical average among all species. Some raise only a couple of litters each year, but others—cottontails, for instance—can shuck out eight or more in a year when the weather is favorable. At the height of it a female may breed when her most recent litter is scarcely a day old, and most of the breeding females in a given population can be pregnant and nursing at the same time. Although most of the young will not breed until the following spring, a substantial number of early-born females will bear litters of their own before the year is out. No doubt this exerts a certain wear and tear on the breeding population, but it certainly is an efficient high-volume system.

At birth a cottontail is about four inches long and weighs an ounce, and four or five at once make a nestful. The female spends her days in a form nearby and visits the nest only to nurse, usually at dawn or dusk. No one knows exactly how many times she does so each day because she's highly secretive about it. We do know that cottontails nurse by squatting over the nest or, when the young are a bit older, by sitting with her hind legs spread to allow the little ones to reach her nipples.

We also know that cottontail milk is among the most nutritious of all mammals', and the young develop rapidly, growing a full coat of fur and gaining use of their eyes and ears within the first eight days. They leave the nest after about two weeks and are independent by the age of one month.

Nearly half the newborn population won't survive that first month, drowned or chilled in the nest by cold rain, found and eaten by dogs, housecats, blacksnakes, crows, raccoons—you name it. It's a tough world, and not many cottontails live to be a full year old. They're capable of living longer, of course, perhaps as long as ten years, but only a very few, very lucky ones make it to two or three.

Most of the mortality is the result of predation, but rabbits also are susceptible to a whole host of parasites and diseases—notably myxomatosis, which is a mosquito-transmitted viral disease, and tularemia, a tick-borne bacterial disease commonly called rabbit fever.

Myxomatosis originally occurred naturally only in South America, but it was deliberately introduced into Australia to help curb exploding populations of rabbits that had themselves been transplanted to the continent. It worked well enough in Australia, but from there it spread to England in the 1950s and devastated British rabbit populations as well—in all, a textbook lesson in what can happen when animal species are introduced to places where they don't belong.

So far as I know, myxy represents little threat to man, but tularemia is another matter. Besides rabbits and hares, it affects some rodents and other mammals, and even certain birds. Humans can become infected through tick bites and contaminated water, but the most common means is through contact with diseased rabbits. More than a few biologists and hunters have contracted rabbit fever while dissecting specimens and dressing game, typically through a cut on a finger or hand or by being stuck with a splinter of bone, either of which allows the bacteria into the bloodstream. Tularemia occasionally is fatal, and even when it isn't the effects can linger for years. I know a couple of people who've had it, and it sounds to me like no fun at all.

Thanks to some new drugs tularemia can now be treated much more effectively than before, but prevention is still the better course. Fortunately it's relatively easy to avoid. Simply stay away from any rabbit that appears sick or sluggish. Don't hunt them until the weather turns cold, since most diseased animals won't survive the first few hard freezes. Wear rubber gloves when you dress them, and check the livers for spots. Most rabbit livers show blemishes of some sort, but one that is speckled with a multitude of fine, white spots is a sign of rabbit fever. Discard such carcasses in a way that pets and scavengers can't get to it. Finally, cook rabbit meat thoroughly.

Statistically the likelihood of contracting tularemia is about the same as the risk of being run over by a buffalo, but still several dozen cases are documented every year. Considering the number of cottontails I dressed bare-handed before I learned enough about tularemia to take it seriously, I suspect Dad's dicta on not hunting rabbits until snowfall and only shooting the ones that ran did more than anything else to keep me from becoming a statistic myself.

I haven't hunted rabbits in years, partly because I ate so many as a kid that I lost my taste for them, and partly because I do most of my bird hunting over pointing dogs. There's an old notion among dog men that fooling with rabbits will upset a dog's training. I don't really believe it, but I've always asked my dogs to eschew rabbits as part of the tradition. Sometimes they comply, sometimes not.

But whether I'm hunting birds or just roaming around, I notice that rabbits don't seem as plentiful as I remember them years ago. Not surprising, really, considering that they're poorly equipped to survive in shopping-mall parking lots, heavily grazed pastures, huge grain fields with every square inch tilled up, drenched in herbicides and plowed in the fall; that they find less and less to eat as native flora continues to disappear under a plague of fescue. An animal as prolific as cottontail has no trouble populating the habitat available, but it can hardly live where habitat no longer exists.

Which is not to say that the prospect for rabbits is any grimmer than prospects for the rest of us. I don't see as many as I once did, but I see them nonetheless. There's a goodly number here on the farm, and they take me back to a time that looms sweet in memory—hunting with Dad, catching rabbits in the garden with the box-trap my grandfather made for me, sitting for hours watching the cottontails nibbling clover and cavorting and courting and in general just being themselves.

Occasionally my Brittany girl, October, forgets the dignity of her age and regal bloodlines and goes ripping off after the bouncing flash of a little white tail, yipping like a witless pup. When she regains her senses, I always tell her she should know better, but I don't really mean it. The Creator god may have intended cottontails mainly as someone's lunch, but I doubt he'd be offended to know that they can serve equally well in just making you feel like a kid again.

◆

# 26

# A Rose by Any Other Name

As I mentioned in a prefatory chapter, the scientific names of plants and animals have translatable meanings. Some are rather poetic, others highly pedestrian. Some of the most interesting names are those that offer insight into human attitudes toward certain animals. *Ursus horribilis,* "horrible bear," reveals something about the way we traditionally have thought of the grizzly. So does calling the timber rattlesnake *Crotalus horridus* and naming one of its subspecies *Crotalus horridus horridus,* "horrid, horrid rattler."

And then there are the creatures called *Spilogale putorius* and *Mephitis mephitis*. The former translates as "foul-smelling spotted weasel"; we know it as the eastern spotted skunk. Regarding the latter, technical name for the striped skunk, you have to wonder whether the zoological community is more interested in precise nomenclature or in making sure we get the idea. *Mephitis mephitis* simply means "bad smell, bad smell."

Of the two names *Spilogale* offers the most information. The spotted skunk is not actually a weasel, but it's related to the weasels. All skunks are, and also to the martens, mink, badgers, wolverines, and otters, with whom they are classified as the family Mustelidae.

The family itself is represented on every continent except Australia, but apart from the zorilla and the North African banded weasel, which are similar in shape and coloration, the skunks as we know them live only in the New World. Exactly how many varieties there are has been a matter of some dispute. As recently as the 1930s mammalogists insisted upon no fewer than thirty-two different species and subspecies, many of them distinguished solely by geographical range and minute differences in size and marking. Nowadays most taxonomists seem to agree on eleven species: the striped skunk; hooded skunk; eastern, western, southern, and pygmy spotted skunks; and five species of hog-nosed skunk.

The striped skunk is by far the most familiar and widespread, ranging from coast to coast and from Canada to northern Mexico. Eastern and western spotted skunks cover almost as much territory, from the northern states to Central America. Hooded skunks are rare in this country, found only in the Southwest. Similarly, hog-nosed and eastern hog-nosed skunks range no farther north than the southwestern states. Southern and pygmy spotted skunks are restricted to Mexico and Central America; Amazonian, Andes, and Patagonian skunks, all hog-nosed species, live only in South America.

At least one aspect of skunkdom enjoys universal agreement among scientists and laymen alike: They are strong-smelling animals. Actually every member of the Mustelidae possesses a pair of musk-producing glands—which is one reason why they're considered to be relatives—and none of them is likely to be confused with lilacs or verbena. But apart from the Old World–native polecat, *Mustela putorius*, none generates a musk as potent as the skunks', nor can any deliver it with quite the same verve and finesse. The scent of skunk is so strong and so distinctive that the association long ago passed beyond synecdoche to the point where the smell has literally come to define the creature. As one indication of how far this has gone, skunks once were thought to particularly relish the common plant *Symplocarpus foetidus* simply because its spathes give off a rather skunky odor. The assumption isn't true, but we still call the plant skunk cabbage.

Unfortunately such preoccupation with sheer stink frequently taints any understanding of what skunks are really all about. So let's get it out of the way up front and clear the air, so to speak, for a more sympathetic view.

Fact is, both the musk and the distribution apparatus are highly impressive. The substance is produced by two internal glands, each about the size of a grape, located at the base of the animal's tail. Ducts lead from these to a pair of tiny, nipplelike nozzles just inside the anal opening. To expose them a skunk needs only to raise its tail, and it cocks its weapons by humping its back to gain purchase on the muscles surrounding the scent glands.

With guns thus drawn, what happens next is entirely up to the skunk. Its control is both voluntary and exquisitely fine. It can aim the nozzles independently, fire one or the other, alternate between them, or touch off both barrels at once. By adjusting the nozzles, choosing between them, or twisting its body a skunk can direct its musk straight back, up, down, to either side, or dead ahead. The spray itself behaves like any liquid ejected under pressure, forming a thin stream that can reach eight or ten feet. Accuracy at that distance is remarkable, and on a windless day a particularly angry or upset skunk can fire a stream almost twenty feet and still hit close enough to count. Sooner or later, depending upon the force behind it, the stream breaks down to a mist that's about equally effective.

A skunk rarely empties its arsenal at a single blast, and it's capable of firing as many as a half-dozen shots before the supply of musk is depleted. As it produces only about a third of an ounce per week, and as the scent glands can hold only about a tablespoonful between them, this says something both for the precision of the animal's control and for the potency of the substance itself.

The exact chemical composition varies slightly from species to species, but skunk musk is essentially a sulphur compound, *n*-butyl mercaptan—a thick, oily liquid that may be white, yellowish, or faintly green and is phosphorescent in the dark. Curiously, the spotted skunk, which is the smallest of the species common to the United States, produces the strongest, most noisome musk. But as every dog owner is probably aware, essence of striped skunk hangs on considerably longer, recurring periodically for no apparent reason.

Actually there's a very good reason, and it has to do with thioacetates, chemical compounds that release thiols, which in turn are compounds that produce the stink. In the spotted skunk, musk thiols dissipate freely into the atmosphere; in the chemistry of striped skunks, however, the thioacetates break down more slowly, triggered especially by contact with water. Even a high level of humidity in the air is enough to liberate thiols for as long as two or three weeks after you or your dog has been nailed by a striped skunk.

These lingering remnants are mere love pats, though, compared with the initial blast. At full strength, skunk musk is the olfactory equivalent of a lightning bolt. It makes your eyes burn, swells the lining of your nasal passages, and in some people triggers instant nausea. Taking a charge full in the face, which is where the little buggers like to aim, is enough to lay anybody low, man or beast. The musk once was thought capable of inducing permanent blindness; it isn't but it certainly interferes with vision temporarily, to say nothing of the fact that it hurts like the veriest hell.

Not surprisingly skunks found a niche in folk medicine, probably brought there by the notion that anything foul must be good for something, and if it's really foul it's probably medicinal. I am mercifully too young to have been slathered with skunk fat as a cure for the common cold, but my mother told me of attending classes just after World War I at the Bumblebee School in Jefferson County, Iowa, at times when the one-room edifice was redolent with the scent.

And not everyone finds the smell repulsive. Some seek it out—either on their own or via membership in Whiffy's Club, a Chicago-based organization for people who find the aroma

particularly pleasant. According to what I read in *National Wildlife,* members are provided a source for vials of dilute skunk scent put up in packaging handy for sniffing whenever a devotee feels the urge.

Obnoxious as it is to most, musk actually has given skunks a bum rap, because despite packing some of the most effective weaponry in nature—or perhaps because of it—they are thoroughly amiable little creatures, playful and gentle and willing to get along peacefully so long as they aren't cornered or seriously threatened. Even then they are remarkably long-fused, more likely to negotiate than slap leather even under fairly extreme circumstances.

◆

A skunk that feels put upon typically sends a warning, often more than one. It might growl or hiss or snap its teeth or stamp the ground with its forepaws. This latter move is very rapid, surprisingly loud, and can be highly amusing if you take it seriously and back off from the encounter. As another little admonitory trick, both spotted and striped skunks sometimes perform a handstand, rear up on their front legs with their tails held high, perhaps even handwalk a few steps. This, too, is fun to watch, but here again observation is best accompanied by retreat.

As its most common cautionary signal a skunk simply raises its tail with the hairs erected at right angles, so it looks like it's wearing a fat, black bottle brush. One old wives' tale has it that the animal doesn't really mean business unless the very tip of the tail is elevated as well; I don't know how this got started, but I can tell you it came from some old wife who never spent much time around skunks.

No matter what sort of warning it gives (and a few grouchy specimens offer none at all), the countdown to blastoff usually ends with the skunk in a U-shaped posture pointing both its nose and its butt your way. When things reach that stage you're about to get whacked.

Nonetheless you stand a better chance of being stung by a bee or bitten by a snake, even where skunks are locally abundant. (This does not apply to dogs, but I'll get to that later.) They really are peaceable chaps, for one thing, and for another they're mainly nocturnal, abroad in late afternoon or evening and on through the night, going about their business with a live-and-let-live attitude that's both admirable and fortunate for us.

Except during the breeding season, which occurs in late winter, a skunk's usual business is finding something to eat. Hearing and smell are its most useful tools in this pursuit, aided by a truly catholic taste. Few mammals are more happily omnivorous. Animal or vegetable, it's pretty much all the same to a skunk, although the proportions naturally vary from season to season. Besides a whole array of fruits, buds, grasses, grains, and assorted plant matter they like mice, rats, and other small rodents, nestling birds, moles, shrews, and almost any sort of carrion. They really don't seek out eggs but they're opportunists so they occasionally get crossways with poultry farmers.

Mainly, however, skunks prey on insects—everything from adult beetles, grasshoppers, and crickets to all manner of grubs and larvae they dig up with their long front claws. Spotted skunks eat proportionately more insects than their striped cousins do, but one Canadian study estimates that striped skunks consumed upwards of 115,000 white grubs in one eight-acre tract over just a few weeks' time.

Perhaps predictably enough, not many predators go looking to turn the tables, but almost any—be it coyote, fox, badger, bobcat, great horned owl, feral housecat, or cougar—will gamble griefs with a skunk if pressed by starvation. Since birds can attack from the air and have virtually no sense of smell, great horned owls probably kill more than any other predator, but even at that I suspect more skunks die refusing to give ground to automobiles than are taken by all the predators together.

For most of the year skunks are not enthusiastic travelers, content instead to rummage around home ranges that may cover no more than a quarter-mile-square for a spotted skunk or a mile and a half in diameter for a striped skunk. Males tend to be more mobile than females anyway, but the onset of mating season sends them packing, perhaps to cover four or five miles in a single night in their characteristically rolling, somewhat waddlesome gait. If he finds a receptive female he'll tarry awhile and then be off again. Not surprisingly the boys get a bit loony during this period, quarrelsome and belligerent enough to bite without provocation or fire their musk at anything that might seem like competition—be it another male, a cow, even a rock or tree whose intentions appear suspect. Their former lady-friends are better off without their company.

As I explained in an earlier chapter, the embryos of a few mustelid species—like those of some other, unrelated animals—

undergo a curious hiatus in their development, a phenomenon known as embryonic diapause or delayed implantation. Eventually something prompts them to implant in the uterus and resume their development until birth.

In any case, delayed implantation does occur in some skunks. The gestation period for striped skunks typically is sixty-two to sixty-six days but may be as long as seventy-five days. Embryonic diapause definitely occurs among western spotted skunks, who breed in late summer but do not bear young until April and May; it has not been fully proven in the eastern species, for whom gestation is a fifty- to sixty-five-day affair.

In preparation for giving birth a female may excavate a simple burrow, but she's more likely to take over more elaborate digs built and subsequently abandoned by some other animal, a woodchuck, muskrat, or fox. The young—five on the average for both spotted and striped skunks—are born blind, deaf, toothless, and virtually naked, although their species' characteristic patterns of black and white markings show clearly, as if they were born elaborately tattooed. They will sport coats of fuzz in a week and will be fully covered with short, soft hair a week after that. Striped skunk babies open their eyes toward the end of the third week; spotted skunks won't open theirs until about thirty-two days. At about three weeks, however, the young of both species are able to assume some semblance of the tail-raised defensive posture, although their glands won't actually produce any musk until they're about six weeks old.

Early on, a nest of skunks is a continual chorus of chirping and pleasant, high-pitched squeaks. Then the character changes from birdlike to kittenish, for young skunks are as playful as any animals alive. And they never entirely lose their taste for fun, which is one reason why skunks, especially striped skunks, make reasonably good pets.

A female with a newly weaned litter hasn't much time for recreation. From her perspective the babies are a cadre of appetites and curiosity, demanding ceaseless feeding and a watchful eye. As if that weren't enough she also has to be on the lookout for any males that might wander by, because once their hormones subside they aren't the least bit shy about eating their own offspring.

By three months the young ones are able to accompany their mother on her nightly foragings, typically trailing along behind her like a string of uniformed recruits. This is an important training period for them as they begin to discover what the world has to offer by way of skunk victuals, where to find it, and how to catch or dig it up.

Although basically friendly animals, skunks are not gregarious in the sense of choosing to live in close contact with others of their kind. The family group is their only social unit, and apron strings begin to loosen in the fall. By the time winter comes on, young and adults alike will have gone their separate ways.

In the northern latitudes winter naturally thins out the available bill of fare. Faced with reduced circumstances skunks typically find a burrow, furnish it with dry leaves, and drop off to sleep. Oftentimes they end up sharing multichambered burrows with woodchucks, possums, or rabbits, and in wickedly cold weather several skunks may snuggle into the same den for communal warmth. They do not truly hibernate even in the depth of winter, but they're perfectly capable of snoozing for several months at a stretch. If a warm spell comes along they're apt to bestir themselves, find a snack, and then trundle back to bed.

For all the skunk's agreeable nature humans still have cause to be wary, and not just because of the musk. More than half the cases of rabies documented in the United States each year are found in skunks. No one knows exactly why, but the most likely explanation centers ironically upon the very effectiveness of their defenses. Skunks are supremely confident of their ability to ward off attack; they'll back away, but they won't back down. Unfortunately, another animal in the advanced stage of rabies is too far gone to care if it gets sprayed. Rabies does not progress in skunks the same way it does in dogs and cats, so the virus may be present in the skunk's saliva—and therefore communicable—well before the animal shows any symptoms.

This has been exactly the case in some well-publicized instances where a single pet skunk has infected as many as twenty-nine people before anyone had a clue it was rabid. Moreover, the standard vaccine so effective in inoculating dogs and cats doesn't necessarily immunize skunks, which has prompted several states to declare pet skunks illegal regardless of how they are obtained.

Statistically, however, you stand about as much chance of catching gonorrhea from a Saint Bernard as you do of getting rabies from a wild skunk. The disease certainly is present, but it rarely is transmitted to humans. Only thirteen human cases of rabies have been documented over the past ten years. Mary Jane Schmidt of the National Centers for Disease Control tells me that only one American contracted rabies in 1990, three in 1991, and one in 1992. None came from skunks; three were infected by bats, one by a dog, and the other contracted the disease from an unknown source in India. The CDC hasn't documented a single instance of a human getting rabies from a skunk since 1981.

The best thing you can do with skunks is simply keep a respectful distance—which isn't a bad idea under any circumstances—and don't worry about rabies if you're merely sprayed. The virus has never been found in musk. Remember, too, a skunk that appears unafraid isn't necessarily rabid; skunks aren't much afraid of anything. And as I mentioned earlier males in the grips of mating frenzy are apt to do all sorts of weird things, but it isn't rabies they have on their minds. Finally, it should go without saying (but unfortunately doesn't) that your dogs and cats ought to be vaccinated and the boosters kept current.

Although dogs stand a greater risk of contracting rabies than people do, it's minuscule compared with their odds of being sprayed. Most dogs that get waxed with a good shot to the face usually do not become repeat offenders, but if it doesn't suffer the pain of musk in its eyes, an aggressive or overly curious dog can persist in fooling with skunks, because the smell alone doesn't seem to put them off. So sooner or later just about every dog owner will face one of the great dilemmas of life in North America—how to get rid of that God-awful stench.

The suggested antidotes make up a list as long as your arm—gasoline, benzine, ammonia, sodium hypochlorite, chloride of lime, oils of citronella and bergamot, turpentine, lye soap, chloroform, you name it. Tomato juice is the classic, but in my experience tomato juice does a lot more good mixed with liberal portions of vodka and taken internally while you treat the stink with something else. I've tried most of the common cures as well as a few exotics, but vinegar seems to work best of all.

I've never told this story in print before (you'll presently see why), but it happened just a couple of months after we moved to our farm in the Ozarks. I got up that day at my usual time, about five in the morning, and as usual my Brittany girl, October, got up, too. I let her out in the frosty November dark, and two or three minutes later, right on cue, she barked at the back door. When I opened it she breezed in wearing an aura of *Mephitis* so nearly palpable that I almost went face-down on the mudroom floor.

Now, I know the scent of skunk. I've smelled it all my life, been sprayed myself, and have tried to bathe it off dogs more times than I can count. I don't find it particularly offensive in small doses; it reminds me of burnt coffee. I'd recognize it anywhere. Right at that moment, however, all my synapses slammed shut, and the only thought coming through was that the propane furnace was leaking gas. (I've lived in the country for years but never before with enough propane in the back yard to level a small village, and I was a little goosey about it at first.)

So instead of sending Tober back outside I flung open the furnace-room door and got down on the floor to see if the pilot light was burning. Since we've been virtually inseparable all her life, my old puppy takes a keen interest in whatever I do, so she got right down there with me to have a look for herself. Then I noticed three things all at once: The pilot light was going strong, Tobe's normally snow-white chest was an oily-looking pastel yellow, and as it was eight inches from my nose right then I was about to pass out from lack of oxygen.

Just as I finally began to sort things out Tober decided that whatever I was doing in the furnace room wasn't going to produce any fun, ran to the other end of the house, and jumped in bed with my stepdaughter, who woke instantly and with every bit as much aplomb as you'd expect from a twelve-year-old being asphyxiated. By the time I gathered my remaining wits, everybody was awake and Tobe was wondering why nobody liked her all of a sudden. I suspect she was equally bemused at being taken outside and drenched with white vinegar, but given her lifelong affinity for getting dirty, stinky, or preferably both, she humors me in these things. She's had some other skunk skirmishes since, but they've been mere nicks by comparison.

The key to using vinegar, I've found, is to not be stingy with it. For the first go-round, pour it on full strength, work it in until the dog's coat is saturated, and let it dry without rinsing. If you follow up later with a bath, a stout mixture of vinegar and water used as a rinse (and left to air-dry) will help control the residual thiols released by the bathwater. You can keep some in a spray bottle for occasional touch-ups later.

Just a few weeks before this book went to the printer, a reader sent me a newspaper clipping about a biologist in West Virginia who proved that skunk musk can be neutralized with a simple mixture devised by a laboratory researcher in Illinois for scrubbing hydrogen sulfide from waste-gas streams. It's called Krebaum's Formula, after the researcher, and it's simply a quart of 3 percent hydrogen peroxide fortified with a quarter-cup of baking soda and a teaspoon of liquid soap—stuff you can buy at any supermarket or drugstore.

Mercifully, I haven't yet had an occasion to use it, but I'm assured it works, and it makes perfect sense chemically. Although two different studies have turned up no harmful side effects, it is so far a pets-only treatment; just keep it away from your dog's or cat's eyes, nose, and mouth, and thoroughly rinse the animal with clean water afterwards.

Vinegar also works well at removing *eau de putois d'Amérique* from clothing, carpets, and furniture. The need to fumigate your living space may not seem very likely, but never say never. I have a friend whose Irish setter once caught an uninjured, highly indignant skunk and paraded it all through the house seeking praise for being a crack retriever. I have related this to Tober, who is an exceptional retriever herself, along with an assurance that I will not be amused by a similar feat on her part. She has agreed, in principle at least, but there's always a gallon of vinegar in the pantry. Skunks are plentiful here, so I figure it's just a matter of time.

# VI

# Grace and Glory

# 27

# Feathered Glory

*Unwearied still, lover by lover,*
*They paddle in the cold*
*Companionable streams or climb the air.*

William Butler Yeats
"The Wild Swans at Coole"

Zeus—cloud-gatherer, lord of the sky, and chief god of Greek mythology—was not always a companionable lover, certainly not on the day he spied Leda, wife of the Spartan king Tyndareus, bathing in the river Eurotas. On impulse, lusty Zeus transformed himself into a swan and ravished her. The aftermath of "that white rush," as Yeats describes it in his poem "Leda and the Swan," would transform the world. From their union was born the lovely Helen, destined to be abducted by the faithless Paris and thereby become the object of contention in a war that left the great ancient city of Troy in ruins.

Although the Ledaean myth certainly involves more dire consequences than any other appearance of swans in folklore, the Greeks were by no means the first to appreciate their elegant beauty and sheer brute power—assuming of course that the swanlike forms of Stone Age cave paintings are what they appear to be. The birds have been subjects of legend, fairy tales, art, literature, music, and wonder ever since. Man it seems has always responded to the swan, not always kindly nor always wisely when kindly, but certainly creatively.

Swans are the largest of all the waterfowl. Taxonomically they're classified with geese and ducks in the family Anatadae, with geese in the subfamily Anserinae, and by themselves in the tribe Cygnini. Six species occur in various parts of the world. Three live in North America: *Cygnus buccinator,* the trumpeter swan; C. *columbianus,* the tundra swan; and C. *olor,* the mute swan. Trumpeters and tundras are native here. The mute is a semidomesticated, Old World transplant.

Until about ten years ago the tundra swan was called whistling swan. Why is anyone's guess, because C. *columbianus* makes no sound that even remotely resembles a whistle. But then none of our three swans vocalizes in ways that fit its common name. Although usually silent, mute swans are anything but mute, and a trumpeter sounds more like a French horn. My friend Ed Zern, exercising his famous wit, once summed it up this way: "The Mute Swan is so called because it makes a loud trumpetlike noise. The Whistling Swan is so called because it barks and whoops. The Trumpeter Swan groans. This is called ornithology."

The species aren't always easy to identify by sight, either, at least on first glance. They are all big birds, extremely long-necked and dressed in snowy white plumage as adults. Only a closer look—and some non-tongue-in-cheek ornithology—brings out the differences.

Mute swans are easiest to identify. On the water, they hold their necks in the classically elegant S-curve that has come to be the archetypal swan profile, their bills pointed slightly downward. In swimming posture, their secondary wingfeathers stick up above the level of their backs. Trumpeters and tundra swans hold their necks straight up and their bills level.

Heads and bills are the most telling characteristics. An adult mute swan's bill is deep reddish orange, black at the tip and base, and shows a distinctive black knob where the bill meets the forehead. A tundra swan's bill is all black, often marked with a small yellow spot just ahead of the eye, and is concave, which makes the bird look somewhat round-headed. (In the Old World race of tundra swan the bill is yellow at the base. This bird, *Cygnus columbianus bewickii*, or Bewick's swan, occasionally shows up in Alaska and on the Northwest coast.) The trumpeter also has a black bill but less concave than the tundra's, which lends a rather flat-headed profile. The trumpeter's bill also shows a characteristic "grin line"—a narrow, salmon-red stripe along the edges of the lower mandible.

Size is another obvious difference, even if you aren't likely to see all three species together outside a zoo. The tundra swan is the smallest—although at about three feet from beak to tail, eighty inches across the wings, and an average weight of twenty pounds, it still is decidedly larger than all but the biggest giant Canada geese. Mute swans are larger yet, but an adult trumpeter is the behemoth of waterfowl: five feet or more in length with an eight-foot wingspan and a thirty-pound heft. For a more graphic sense of these dimensions, consider that the walls of a typical American home measure exactly eight feet from floor to ceiling, and next time you're in a supermarket pick up three ten-pound bags of potatoes at once. These are big birds.

Most of a swan's overall length is in its neck, and a swan's neck is a remarkable structure. For one thing it comprises more individual bones than the neck of any other warm-blooded animal. Mammals, whether mice, humans, or giraffes, have only seven cervical vertebrae. Pigeons have fourteen, swans twenty-three. Nearly all birds' necks are highly flexible, which compensates for the rigidity of their virtually solid backbones, but none is so sinuous as a swan's.

This pliability combined with sheer length allows them to exploit a feeding niche literally beyond the reach of most other wildfowl. Swans are for the most part vegetarians with a great relish for the roots of wild celery and other succulent aquatic plants, and an eighteen-inch neck is an excellent tool for getting at plants in water too deep for geese and dabbling ducks. If its normal feeding posture—body floating horizontally with head and neck submerged—isn't quite enough, the bird can gain a few more inches by tipping its tail in the air and getting better than half its body underwater as well. Ducks and geese do this, too; it's the anseriform equivalent of a boarding-house reach.

A swan's feeding apparatus is also unusually well protected. All waterfowl wear a thick layer of down under their contour, or outer, feathers, but none can match the garb of a swan. A mallard in full plumage has about 12,000 feathers; the much larger bald eagle has only about 7,100. But in November 1937 biologist Andy Ammann plucked and counted 25,216 contour feathers from a tundra swan, and fully 80 percent of them came from the neck and head—which suggests that no swan is likely to develop a headache or a sore throat from foraging in even the coldest water.

Plants aren't the only edible stuff to be found on the bottom of a lake, river, pond, marsh, or estuary, and swans have no aversion to supplementing their diets with a quotient of animal matter—worms, insects, thin-shelled clams and mussels, snails, tadpoles, frogs, even small fish. In areas where soil erosion, agricultural runoff, and pollution from other sources increases water turbidity, blocks sunlight, and thereby reduces the amount of aquatic vegetation, the birds have proven adaptable enough to prosper in corn and beanfields, stands of winter wheat, cranberry bogs, and pastures.

Regardless of what they're eating, swans have hearty appetites and typically put away five to ten pounds of food per adult bird every day. In places where natural foods are scarce and swans aren't, this can get them in dutch with the farming community. Although their agricultural depredations can be severe, it isn't a widespread problem, mainly because swans are not so plentiful as they once were.

◆

The first European settlers to reach the New World found tundra swans in great numbers along the wintering grounds of the middle Atlantic coast. Those who explored the interior saw them along the Ohio and lower Mississippi rivers and on the Texas and Louisiana Gulf coasts. No one knew it at the time, but both native species ranged over much of the northern United States and Canada. Trumpeters were more plentiful west of the Great Lakes and Hudson Bay; in colonial times they nested all across the upper prairie regions, reaching as far south as Missouri and the northern two-thirds of Illinois.

Not surprisingly North American swans shared the brunt of the nineteenth century's cavalier consumerism. Their feathers ended up in pillows and comforters and as decoration for clothing and hats; their down became powder puffs, their wing-feathers quill pens; their skins went to a variety of uses and their bodies to table fare. (Young swans aren't bad eating if you're fond of strong-tasting wildfowl flesh, but take it from me, an old bird is every bit as tasty as a haunch of radial tire and about twice as tough.) From 1853 to 1877 the Hudson's Bay Company sold 17,671 swan skins. A feather merchant in the eastern United States marketed more than 108,000 skins, most of them trumpeters, between 1820 and 1880.

After that the market fell apart, mainly because there were few swans left to harvest. Their last recorded nestings came from Idaho in 1877, from Iowa in 1883, from Minnesota in 1886, and from North Dakota at about 1890. No nests were reported even in Alberta for many years after 1891. By the turn of the century trumpeters were not known to nest anywhere in the United States, with the possible exception of a few holdouts in Montana. Even the sight of one south of the Canadian border was an item of news and note.

The situation north of the border wasn't much better. Native peoples plundered the trumpeters' nests for eggs and killed the birds for food, running them down on foot during the molt, when like all waterfowl they are flightless. This was part of an age-old relationship between the tundra-dwellers and the swans, but now, beset from all sides, the birds could ill afford the losses.

In September 1907 Ernest Thompson Seton saw a dozen small flocks, totalling about a hundred birds, near the mouth of the Mackenzie River on Great Slave Lake in the Northwest Territories and wrote, "It rejoiced my heart to see even that many. I had feared that the species was far gone on the trail of the Passenger Pigeon."

Most others shared the same fear. Writing in 1912, naturalist E. H. Forbush predicted the trumpeter's fate this way: "Its total extinction is now only a matter of years...In the ages to come, like the call of the whooping crane, they will be locked in the silence of the past."

As it happened, Forbush was wrong—but not by much. The Migratory Bird Treaty Act of 1918 extended full protection to all swans, but by 1931 only thirty-five trumpeters were known to exist in the United States, most of them living around Yellowstone National Park. The count rose slightly, to forty-six adults, in 1935, the year President Roosevelt established Red Rock Lakes National Wildlife Refuge in Montana's Centennial Valley, just west of Yellowstone. (Another flock of seventy-seven adults and assorted young were then living along the eastern slope of the Canadian Rockies near Grande Prairie, Alberta, but no one outside the local area knew about them until some time later.)

With ponds fed by the same warm-water springs that plumb Yellowstone, the Red Rock Lakes area proved an ideal nursery from which the trumpeter swan population could be reborn. Some ornithologists feared their numbers had dropped so low as to wipe out the species' genetic memory of their ancient migratory routes, but this does not appear to have been the case. It also is likely that there were more trumpeters alive in the 1930s than anyone knew, particularly in view of the fact that a flock of several hundred was discovered in 1954, keeping company with tundra swans on the Copper River Delta in Alaska.

◆

The trumpeter swan returned in any event. The population in Canada and Alaska was up to about 1,500 by 1960. An Alaskan survey showed about 2,800 in 1968, and now they number more than 15,000 continent-wide. Alaska still has the most, known as the Pacific Coast population, but the Rocky Mountain population is nearly 3,000 strong, and several hundred now live in the heartland, thanks to releases at federal refuges throughout the Midwest over the past twenty years.

In contrast to the trumpeter's history of bust and boom, tundra swans managed to survive the bad old days with their populations relatively intact. Even in the mid-1910s good numbers of them continued to visit the traditional wintering grounds from Chesapeake Bay to Currituck Sound, populations not only holding their own but in some years seeming on the increase. Sixty years later the increases were both apparent and real. By 1975 the eastern population was estimated at 60,000, and the number of tundra swans showing up at the Mattamuskeet refuge in North Carolina had increased from about 5,000 to nearly 20,000 over the previous five years. (Unlike other wildfowl, which use four distinct flyways across North America, tundra swans essentially use only two—the Atlantic Flyway, with wintering grounds on the mid-Atlantic coast, and the Western Flyway, whose travelers winter in Utah and on the Pacific coast.) Now the North American population has swelled to 200,000 or more, enough birds that a limited, carefully monitored hunting season for tundra swans is in force in both flyways—in Utah, Nevada, Montana, Alaska, and the Dakotas in the West, in North Carolina and Virginia in the East.

Ironically, most of what we know about the lives of both our native swans is relatively new knowledge, learned in just the past few decades. In the trumpeter's case the irony is especially pungent, considering that the species reached and rebounded from the brink of extinction by human cause before we had any systematic understanding of the animal itself. But this certainly is not unique; much the same thing happened to the buffalo and the elk and the white-tailed deer and a dozen or more others in whose behalf we smartened up just in the nick of time.

Some of the facts have been clear all along. Swans are migratory birds that nest in the Arctic wilderness and spend their winters in more temperate climes. But until about twenty years ago no one truly appreciated just what magnificent voyagers they really are, not until a team of researchers found tundra swans tagged on Chesapeake Bay summering on Alaska's North Slope, 3,500 miles away. Later studies using radio telemetry have shown similarly impressive penchants for travel.

Swans are immensely powerful in flight, and they tend to be among the earliest wildfowl to head north with determination, forming their migratory groups of ten to a hundred, streaming across the sky at thirty-five to fifty miles per hour in V-form skeins, continually changing positions to conserve energy, as geese do.

Probably for the same energy-conserving reason, they also tend to fly at high altitude—3,000 to 10,000 feet—avoiding low-level turbulence by simply flying over it. Some evidence suggests that birds southbound in the fall fly higher than they do on the northward, springtime trek. This seems to have been the case in November 1963 when a turboprop plane cruising over Maryland plowed into a flock of tundra swans at 6,000 feet. In the collision one of the big bodies smashed a horizontal stabilizer, sending the aircraft and all aboard into a deadly crash.

Such mishaps are quite rare. Typically the birds arrive where they're headed without running afoul of anything more bothersome than heavy weather—which can, however, be bothersome in the extreme. Records exist of flocks being forced to land because their wings iced up.

There is a particular urgency about their travels in the spring, for then the birds are driven not only to find their traditional summering grounds but also to create future generations of their kind.

◆

One bit of swan folklore that seems wound on a bobbin of truth is their reputation for monogamy. Both trumpeters and tundra swans appear to mate for life, pairing up at about the age of three. They do not, however, breed until about their fifth year, which is one reason why their populations reached a nadir years ago; the slower an animal reproduces itself, the less able it is to tolerate disruption in its life.

Even those pairs in breeding trim perform an annual courtship ritual, which may have something to do with their ability to stay together in a lifelong match. It may take place on land or on water, and both sexes take part in an elaborate and graceful minuet of stately bows and turns, wings raised and partly extended.

With the formalities complete, nest building is a main order of business. The pen, or female, seems to do most of the work while the cob, or male swan, stays nearby as if to offer engineering advice. Not, however, that she has much need for it. Once at a site she likes—which may be at the edge of a lake or marsh, on an island, or atop a muskrat house—she uses her powerful bill like a steam shovel and her long neck like a crane, piling up moss, grass, rushes, and whatever else is handy until she has a mound perhaps two feet high and five or six feet wide. In the middle she forms a cavity eighteen inches to two feet across, lines it with down, feathers, and other soft stuff, and over the course of several days deposits from four to a half-dozen eggs.

These eggs would land a chicken, or even a Canada goose, in the nearest emergency room. For a tundra swan they average about four inches long and more than three inches in diameter; trumpeter eggs are roughly 10 percent larger still. Incubation time for both species is about five weeks. The cygnets, as young swans are called, emerge dressed in leaden-gray down, and they retain this dusky color throughout the first year of their lives. For at least a week or two the family stays near the nest where the pen broods her charges at night and on rainy days. Otherwise they spend much of their time on the water, the adults foraging food for themselves and the cygnets, and the little ones learning. While the group is well offshore the cygnets are allowed to paddle about largely as they please, even climb onto an adult's back for a rest. But closer in they form up single file behind one parent while the other brings up the rear.

Between them, a pair of adult swans offers their young a considerable degree of security. Besides a powerful bite, swans have unusually large elbow joints in their wings, and they use these massive lumps of bone like knobkerries. They press their attack at a full run, spin sideways at the last moment and lash down with their wings. Driven by muscles capable of sustaining a twenty- or thirty-pound bird in flight, the wing-joints deliver smashing blows, more than enough to kill a fox or coyote or dog or to break human bones. Trumpeters with cygnets have been known to drive full-grown moose into retreat.

The young swans grow quickly. Trumpeter cygnets will weigh nearly twenty pounds by the time they learn to fly, at the age of three to four months. Not surprisingly swans do not have an easy time getting airborne. Like loons and diving ducks, they have to run some distance to get up enough takeoff speed. Learning to fly must be a frightfully laborious job for the young.

Naturally it takes them a while to build up enough strength to fly any great distance, and P. M. Silloway in 1903 reported an observation from Montana of a trumpeter cygnet taking rest breaks on an adult's back: "The cygnet flew only a few feet directly above the elder...The younger swan would fly 50 or 60 yards alone, then drop lightly upon the parent's back to rest, being carried for 50 to 60 yards in this manner; then it would rise upon its own pinions and flap along with the elder bird until it again became weary of its own exertions." The notion of even a full-grown trumpeter being able to carry a burden more than half its own weight is a wee bit hard to swallow, but stranger things have happened.

(One of which is that the clamoring of a thousand-odd tundra swans, forced by heavy clouds to land on the Allegheny River one night in October 1971, prompted police in the Pittsburgh suburbs to mobilize for riot-control in the mistaken belief that what they were hearing was a resurgence of the riots that had broken out two weeks before, when the Pirates won the World Series.)

By the time fall comes to the Arctic, however, the young birds are well up to the task of migrating south, and there is evidence that tundra swans make an epic journey of it, flying as much as 1,300 miles or more nonstop.

With tundra swan populations flourishing and trumpeters drawing ever farther back from the edge of extinction, it seems likely that our most elegant wildfowl will be with us for quite some time to come. But all is not peachy on the swan front—not because of too few birds but rather because of too many.

As I said earlier, the mute swan is an Old World bird, and it was domesticated in Europe at least as early as the twelfth century. In England swans were property of the Crown and could neither be owned nor bred without permission from the Royal Swanherd, who also was responsible for conducting an annual inventory roundup of His (or Her) Majesty's stock. No nobleman's estate was complete without at least one pair of the great white birds gliding elegantly across the ornamental ponds.

Not to be left out of things, upper-crust Americans of the late nineteenth and early twentieth centuries imported mute swans from Britain and Europe to grace their own estates on Long Island, in upstate New York, and elsewhere. Well protected and with nothing more pressing to do than parade their gorgeousness, the birds naturally were fruitful, and as a result feral populations of mute swans now live in New York, Massachusetts, Rhode Island, Connecticut, Pennsylvania, and Maryland. Others, also descendants of ornamental birds, can be found around the Great Lakes, in Yellowstone, in Ontario, and in British Columbia.

As often is the case with animals transplanted to biomes where they do not naturally occur, mute swans are becoming a problem along the Atlantic Coast. Long accustomed to living in proximity to man, mutes require neither the isolation nor the space that the native swans must have for nesting. Consequently the populations are in some areas increasing by 30 to 40 percent every year. When it involves animals as large as mute swans, that kind of growth can breed disaster.

Not for the swans, of course, but for native wildlife and for the environment as well. They nest earlier than native wildfowl and being marvelously ill-tempered they are diligent about driving off what ducks and geese haven't already been displaced by human presence. Should other wildfowl manage to bring off a successful nesting anyway, mute swans haven't the slightest compunction about killing any ducklings or goslings they can catch. Nor, conditioned by well-meaning but ignorant people to expect food from anything on two legs, are they hesitant about attacking anyone who happens to show up unprepared to pay tribute.

The environmental damage they wreak is truly serious. It doesn't take many twenty-five-pound mute swans to eradicate the vegetation from a brackish coastal pond—which in turn destroys the entire chain of aquatic life the vegetation supports and also destroys the pond's ability to act as a natural pollutant filtration system.

More enlightened states—those where the people with political clout accept the judgment of professional wildlife managers and don't cave in to the weeping and gnashing of citizens who don't know the difference between reality and Walt Disney—control mute swan populations by various means. One way is to find their nests and shake the eggs, which addles the contents and effectively prevents any from hatching. The swans will continue incubating the dead clutch until too late for renesting. Another is to pierce each egg, which allows bacteria to invade and kill the embryos. Yet another is simply to shoot any mute swan seen within a certain locale—say, a wildlife management area.

All of these unfortunately are labor intensive and hindered by public pressure from people who see only a lovely animal and not the implications of its presence. In a survey conducted a couple of years ago by Yale University for the U.S. Department of the Interior to determine the most popular animals in America, dogs and horses came in first and second, respectively, and swans finished third. Not native or tundra or trumpeter swans, just "swans." Not a good sign. By every moral right, habitat belongs to the organisms that have evolved within it, and we all, man and animal alike, would be better off if mute swans ceased to exist in North America, or at least if their numbers were kept very small. This would place the species in no jeopardy at all, and it certainly would help save some animals that truly belong here, along with some habitat we really cannot afford to lose.

One of the oldest and most persistent beliefs about swans is that they burst forth in glorious song at the moment of death—presumably because they have been unable to sing at any other time of their lives. The idea is at least as old as Socrates, who according to Plato toasted his disciples with a cup of hemlock and said, "I do not believe swans sing in sorrow. I believe, because they are prophets, they know the good in the other world. I depart life no more dispirited than they."

Incredibly, you can find otherwise well-qualified ornithologists who still buy the notion, but the fact is, the swan song is pure fancy. An expiring swan exhaling through one of the longest windpipes in the avian world might very well make what Shakespeare (whom Samuel Johnson called the Swan of Avon) describes as a melody with a dying fall, but swans do not sing in the face of death. They only die. Which is enough to ask of any creature.

◆

# 28

# The Voyagers

Since time out of mind geese have conjoined the seasons, gathered and stitched the critical seams of the year, autumn to winter, winter to spring. They tell antique tales of passage, write cryptic, shifting messages in hieroglyph across the sky. They chant old songs in ancient tongues, speak syllables of change and eternity, hardship and hope, until in their wild, brittle voices time itself congeals to an essence.

Geese are essence—the essence of wildness, of power on wings, of wariness and mystery and romance, of sheer spectacle in their profusion, appearing like wind out of nowhere to announce the changing seasons and then moving on. Geese may have been the first animals to give man some concept of a world greater than the scale of his own horizons.

Where do they come from? Where do they go? Questions reverberate in both time and space. During the Middle Ages geese were thought to grow from barnacle shells, an idea that seems to have originated with Giraldus Cambrensis, who visited Ireland in 1187 and wrote of seeing geese that "resemble the marsh-geese but are smaller. Being at first gummy excrescences from pine beams floating on the waters, and then enclosed in shells to secure their free growth, they hang by their beaks, like seaweeds attached to timber. Being in the process of time well covered with feathers, they either fall into the water or take their flight in the free air...I have often seen with my own eyes more than a thousand minute embryos of birds of this species on the seashore, hanging from one piece of timber, covered with shells, and already formed."

Whether this was an attempt to explain the origins of migratory geese not seen to breed in southern Europe, or whether it derived from some other notion now hopelessly obscure, the immediate result was not an unhappy one: Since geese were fish instead of fowl, Giraldus says, "bishops and men of religion make no scruple of eating these birds on fasting days," and roast goose long after was permitted on meatless Fridays in many parts of Europe.

Giraldus's legend is still preserved in the common name of *Branta leucopsis*, the barnacle goose, and in the brant's scientific name, *Branta bernicula*.

Others have approached the question of where geese come from with a geographical slant—but without altogether forsaking a sense of romance; witness the snow goose's old scientific name *Chen hyperborea*, "goose from beyond the north wind."

The most systematic thinking on the origins of geese is less fanciful but no less complex. They belong to the family Anatidae, the waterfowl, which comprises 3 subfamilies, 10 tribes, 43 genera, 147 species, and 247 forms. Because all waterfowl share so many important characteristics, ornithologists have had a devilishly rough time deciding what's truly what—duck, goose, or swan. In fact both *duck* and *goose* are so widely and indiscriminately applied that the scientific community recognizes them as no more than substantive names. Technically, true geese and swans are grouped together in the subfamily Anserinae and the tribe Anserini.

Fossil evidence suggests that geese and swans separated from the other waterfowl about 30 million years ago, during the Oligocene, and have evolved on their own ever since. In the process, the geese have grown more diverse, comprising fifteen species compared with six species of swans. Geese are, moreover, much given to developing distinct geographical races and subspecies—so many in fact that scientists have yet to sort them into a definitive paradigm. (Which if nothing else, I take as evidence in support of my own theory: If all the taxonomists in the world were laid end to end, they would not reach a consensus.)

All species are classified in two genera—*Anser*, or gray geese, and *Branta*, or black geese. *Chen* is still used as a generic name by some authorities, particularly Americans. Most European ornithologists have abandoned it in favor of *Anser*.

Though gooselike birds live all over the world, true geese occur only in the Northern Hemisphere—except for the Cape Barren goose of Australia, which some taxonomists classify in a different subfamily and tribe. Some Anserini species, notably the white-fronted goose and brant, occur throughout the northern half of the world; others cover relatively smaller ranges. The bar-headed, red-breasted, and swan goose are mainly Asian. Species living throughout Eurasia include the graylag, which is the progenitor of most domestic geese; the lesser white-front; the barnacle; the bean goose and its subspecies; and the pink-footed goose—which some taxonomists insist is a separate species entirely.

Some North American geese occur over much of the continent while others are more restricted. The emperor goose—a small bird despite its lofty name—hangs out on the Alaskan coast and along the Aleutian chain. The nene lives only on the Hawaiian islands. Ross's goose, a pure white bird with black wingtips, breeds only in a small area of Canada's Northwest Territories at the base of the Boothia Peninsula and on Southampton Island,

and winters in the Central Valley of California. It is the smallest and rarest of all North American species. A few Ross's geese drift down the Mississippi Flyway every year, but if noticed at all they're usually mistaken for small snow geese. Greater white-fronts are fairly common in the West, rarely seen east of the Mississippi.

Snow geese are considerably more plentiful and in some ways more complex. Wherever you find snow geese, distinctively white with wings tipped in black, you'll also find others of identical size with white heads and blue-gray bodies—blue geese, to most. For years taxonomists insisted they were separate species, calling the white ones *Chen hyperborea,* which is a fine name for a bird that nests on the Arctic tundra up at the top of the world. But as if at pains to highlight the distinction between poetry and science, they called the others *Chen caerulescens,* "dark-blue goose."

For decades these latter birds remained a mystery. The first blue-goose nesting grounds weren't discovered until 1929, on the shores of Foxe Basin, west of Baffin Island in the Northwest Territories. This of course helped solidify their separate-species status. In 1961, however, Dr. Graham Cooch conclusively demonstrated that blues and snows are simply different color phases of the same bird. Since then they have shared the same name, first *Chen caerulescens,* now *Anser caerulescens.*

Taxonomic blessings aside, blue geese and snows do show some geographical and behavioral differences. The white birds nest earlier in the year and farther north, and they cover a much wider range than the darker ones. Where their ranges overlap, the two phases freely interbreed, although white males show some reluctance to mate with blue females while white-phase females readily accept blue-phase mates. Interbreeding tends to produce more than twice as many blue-phase goslings as white ones, and some naturalists speculate that the white phase may in time diminish in numbers, genetically subsumed by the blues.

Which doesn't mean that white geese are doomed, though, because there actually are two races of snow geese in North America. The more numerous and widespread race, the one comprising the two color phases, technically is the lesser snow goose, *Anser caerulescens caerulescens.* The greater snow goose, *A. c. atlanticus*, is about half again larger and lives only in eastern North America, nesting in Greenland, on Ellesmere Island, and Baffin Island, and wintering along the Atlantic Seaboard. So far as anyone knows, no blue color phase occurs among greater snow geese nor among Ross's geese.

No one has yet fully explained why two color phases should exist in one race but not in another. No doubt the dark phase originated as a genetic mutation and has survived because the coloration somehow benefits the animal, perhaps as protective camouflage on the nesting grounds. In a larger sense, however, the tendency to evolve races and subspecies, which is unusual among migratory animals, offers some insight into the nature of geese in general.

Subspecies most readily evolve in animal populations geographically isolated from others of their kind. Fish are probably the classic examples. This clearly is not the case with geese, who gather in enormous numbers. But evolution is a complex process influenced by differences in behavior, in reproductive biology, and in life cycles; in these geese and swans differ significantly from other waterfowl.

Ducks live relatively short lives, reach sexual maturity within their first year, often produce a second clutch of eggs if the first nest is destroyed early in the breeding season, and take up with different mates each year—all of which promotes a high degree of genetic uniformity in the general population and at the same time allows new adaptations for survival to spread quickly on a continental scale. So if one mallard should evolve some new and useful way of exploiting its habitat and produces offspring with the same capability, the new adaptation can spread at a geometric rate among vast numbers of mallards.

Geese and swans, on the other hand, live longer than ducks and do not become sexually mature until they're two or three years old. Their pair bonding is relatively permanent; it isn't strictly true that all geese mate for life, but they do tend to stay with the same mates year after year, and often refuse to take another if the first one dies. Moreover, geese prefer to nest in the same area where they were hatched.

Geese lay fewer eggs than ducks and seldom renest if the clutch is lost. The survival rate among goslings is relatively high, but their clannishness and conservative breeding habits—which amount to a high level of inbreeding—ensure that genetic changes evolve slowly and remain confined to relatively small populations. And these eventually become distinct races and subspecies, maintained by lifestyle and behavior rather than geography.

In none is the tendency toward establishing separate races more apparent than among the most magnificent geese of all—the Canadas.

*Branta canadensis* is the classic North American goose, the archetypal "wild goose" of the New World, a bird of extraordinary beauty and grace. It also is a bird of extraordinary diversity, for during historical times fully a dozen subspecies have been identified and described.

All Canada geese present essentially the same appearance—black head and neck with distinctive white cheek patches that join at the throat to form a chinstrap; breast and belly pale gray fading to white behind the legs; back and upper wing surfaces subtly variegated gray-brown. The subspecies show some minor variations, mainly lighter or darker shades of the same basic colors, but what really sets the various races apart is size. The cackling Canada, which weighs about three pounds on the average, is the smallest. At the other end of the scale a gander of the giant, or *maxima*, race might weigh eighteen pounds or more and own a wingspread nearly six feet across. Between these extremes the other races, in ascending order of size, are the Richardson's Canada, the Aleutian, Taverner's, lesser, Atlantic, interior, dusky, Vancouver, and western.

Count them and they add up to eleven. The Bering Canada goose, a small race that bred on Commander and Kurile islands in the Bering Sea, went extinct shortly after the turn of the century, a victim of human presence and predatory animals that man brought to the islands. The Aleutian Canada, which once bred all along the Aleutian chain, nearly suffered the same fate.

And then there's the giant Canada. For many years, some naturalists considered it extinct, while others sincerely doubted that it had ever existed. Lewis and Clark wrote of seeing huge geese along the Missouri River a hundred miles or so upstream from St. Louis. Later in the nineteenth century and into the twentieth, old hunters in western Iowa, eastern Nebraska, and the Dakotas spoke of shooting eighteen- to twenty-pound honkers—reports that many ornithologists simply dismissed as old-timers' tales. A few, like James Moffit and Jean Delacour, were more inclined to believe that a giant race of geese had once existed.

In 1951 Delacour gave the race the name it still bears, *Branta canadensis maxima*, and in 1954 wrote that the giant Canada "appears to be extinct."

Nearly everyone with any scientific credentials at all agreed, at least until 1962 when Harold Hanson, a biologist with the Illinois Natural History Survey, found a flock of giant Canadas living happily and well on Silver Lake in a residential section of Rochester, Minnesota.

The discovery, announced after a careful study in which Hanson established the birds' identity and determined an overall population of about 55,000, elicited a great stir among waterfowl biologists but little more than a collective shrug from the citizens of Rochester, to whom the big geese had been a familiar sight for years. As a boy David Maass lived about two blocks from Silver Lake. "It was my main reference area," Dave told me. "I can't tell you how many drawings I made of giant Canada geese before I, or anyone else, knew what they really were."

Once resurrected from presumed extinction (or in some quarters from the realm of myth), giants began to show up elsewhere—or rather were recognized in places they'd been all along, including the Missouri River bluffs where Lewis and Clark had seen them in the first place.

Rumors that geese nested in the bluffs had floated up and down the river for years, but in 1971 a farmer near Hermann, Missouri, reported actually seeing a nest. Wildlife biologists Charles Schwartz and Glenn Chambers investigated by helicopter and came back with photographs that proved the farmer right. David Graber, a graduate student at the University of Missouri-Columbia, began a thorough study soon after, searching a hundred miles of river bluffs between Jefferson City and St. Charles. Frequently he rappelled down the sheer, crumbling limestone faces, some of them 200 feet high, to check likely sites or to examine nests located by helicopter reconnaissance. He found thirty-two nests in all—on terraces, ledges, and outcrops, in crevices, and among the cedars growing at the blufftops—and he discovered that these cliff-dwelling birds, one of the most unique populations of geese anywhere, were giant Canadas.

What makes them unique is not that they breed in these southern latitudes but rather that they managed to continue doing so largely unnoticed and unaffected by the presence of man. Although the majority of geese worldwide nest in the Arctic or subarctic, vast numbers of Canadas traditionally chose their breeding grounds farther south, along the mid-Atlantic coast and all across the midwestern prairies and the Great Plains. Human settlement and exploitation eventually pushed these populations out. Now they're coming back.

Ironically man himself has played a key role in rebuilding the southern-breeding flocks. For the habitat we have destroyed, we've created some, too. Reservoirs large enough to remain at least partially ice-free are good wintering grounds. So are rivers and lakes where warm-water discharge from power plants provides open water year-round. Artificial nests have proved wonderfully successful; if properly placed in even a small farm pond, a galvanized washtub slightly modified and mounted on a post works beautifully.

Agriculture provides a cornucopia. Geese are grazers, grass- and grain-eaters of great appetite, and love few things better than a field of winter wheat or sprouting corn or beans or oats. They're fond of aquatic vegetation, too, but they can dine as happily on golf courses, city parks, and suburban lawns. In the fall they glean cut grainfields.

None of this would work nearly so well if the birds were unable to adapt, but a Canada goose is a consummate opportunist. While snow geese persistently cling to their Arctic breeding grounds, Canadas are inclined to believe that going far enough is just as good as going all the way. The flock that winters at Rochester, for instance, nests on the western side of the Delta Marsh in southern Manitoba, a straight-line distance of less than 600 miles. Other flocks cover even less ground in the course of a year, nesting wherever the habitat is suitable, and flying only far enough south to find open water and something to eat.

Wherever they nest, whether the Arctic tundra, the northern prairie, or farmland farther south, the actual process is much the same. Come spring, the mysterious and complex mechanisms of migratory imperative sends the voyagers on their way.

◆

Formation flying is characteristic of all geese. It's mainly an energy-saving device, as each bird breaks a trail through the air for the one behind, and they change places frequently, so the same birds neither lead nor follow all the time.

Even when they're only distant specks against the sky you can tell who you're seeing if you know what to look for. Snow geese fly in long, undulating lines and shallow V-formations, Canadas in wider, more sharply defined vees that move straight ahead. Snows migrate in much larger flocks than Canadas do and usually begin their flights at sundown. Canadas seem willing to travel at any time of day. All geese talk among themselves as they fly, but Canadas are compulsive about it, constantly calling back and forth as if no detail were too trivial to go unremarked.

Sun, moon, stars, wind, and the earth's magnetic field all seem to play certain roles in how geese navigate, but this, too, is a phenomenon that so far has accrued more questions than answers. Somehow they find their way and arrive at the breeding grounds sometime between early March and the end of May, depending on how far north they go.

All waterfowl perform complex courtship and breeding displays, and each species has its own unique ritual for establishing mates. Among Canada geese, males go a'courting by swimming about near the females, striking haughty poses, and generally advertising what lordly specimens they are. Pair-bonding seems to be accomplished by a ritual known as the "triumph ceremony," which is usually performed after the male has defended his territory by threatening or attacking another goose. He approaches the female with his neck stretched forward, honking loudly. She does the same until they're facing one another, necks and heads side by side, shouting in each other's ears at the top of their lungs. The whole thing bears little resemblance to the whispering of sweet nothings, but it has the same effect.

Canadas perform yet other rituals around the mating act, which takes place on the water. The prelude involves elaborate, mutual head-bobbing and dipping. Afterwards they call loudly, necks outstretched and wings partially or fully raised.

Unlike snow geese, who nest in great colonies, Canadas insist upon some space, as much as two hundred yards or more from their nearest neighbors. The nest itself is a casual affair cobbled together from whatever is handy. The female lays her first egg within about ten days after the nest was begun and adds another every second day until she has a clutch of four to six, cushioned in down that she has plucked from her lower breast.

Incubation takes from three weeks to a month according to species, but once hatching begins the entire brood emerges within twenty-four hours. Thanks to reserves of body fat they developed while in the egg, the goslings can live several days without food, and thanks to their parents' attentive care they aren't particularly likely to be snatched by some predator. But their first day of life is a critical time nonetheless, for this is when imprinting takes place.

Through some mechanism, newly hatched waterfowl quickly become conditioned to recognize and subsequently follow the major object in their environment. In most instances, of course, they imprint on their parents, which certainly improves their chances of survival and is moreover one key reason why waterfowl are gregarious. But as experiments with artificially hatched birds have proven, they will imprint on anything—a human, a dog, a chicken, a rag doll, or a bucket—and goslings imprint more readily than ducklings, so immediate contact with the parents is extremely important.

The process may actually begin earlier. In the last few days before they hatch the young begin peeping, and the female mutters softly in response. The goslings may thus make auditory associations with their mother and with one another even before they leave the eggs.

In any event the female usually stays on the nest for a day after the brood hatches, but on the second morning she leads them out into the world, a matter that may be as simple as walking away or as hair-raising as skydiving without a parachute. All goslings leave their nests no matter where they are, and those hatched on river-bluff ledges simply jump. A day-old *maxima* gosling might weigh three ounces or more, nearly the heft of a baseball, and the drop from the nest to the stony rubble below may be ten yards or 150 feet. Their free-fall might be cushioned by treetops or brush, or it might not. That any survive is a wonder; that most do is, as my old friend John Madson puts it, a small miracle.

Geese are by nature somewhat less aquatic than ducks, but open water is a vital component of brood habitat. The trip from nest to water may be an easy stroll of a few feet or it may be several thousand grueling yards over rocks and logs, railroad tracks and highways, through tall grass, brush, willow slaps, or an immense expanse of tundra. All the while they're exposed to predators both land-bound and on the wing. The goslings that survive to reach the brooding area still remain vulnerable to the attentions of hawks, owls, gulls, turtles, predatory fish, foxes, coyotes, raccoons, dogs, cats, and almost everything else with a taste for fowl.

They are not, however, without a measure of defense. Surprised by a predator on land, the little geese scatter like quail chicks. On water even day-old goslings can dive with remarkable skill. Their parents possess extremely keen eyesight and hearing and are decidedly short-fused with intruders. They can't do much about attacks from the air or underwater, but anything that approaches the goslings by land is apt to be met by powerful bills, sharply clawed feet, and thrashing wings strong enough to break a man's arm.

During the first days of their lives the goslings stick close to their parents. You can, in fact, get some idea of how old the brood is by the way they swim. Little ones paddling along in a tight cluster behind one of the adults are less than a week out of the egg; after a few days they begin to venture away from the parents a bit, and when being led often swim strung out in a line.

◆

They gain size even more rapidly than confidence, and typically double their weight in the first week. In another week they may weigh a pound or more. At about that time their olive-yellow natal down begins to turn a nondescript shade of locker-room gray, and their first suit of grown-up feathers starts growing. At about two months they're nearly the size of their parents.

The adults meanwhile are growing some new feathers of their own. All birds molt at least once each year. It's a gradual process for most, as new feathers replace old ones more or less at random. Waterfowl, however, molt all their major wing- and tail-feathers at roughly the same time and are flightless for four or five weeks until new ones develop.

Nonbreeding geese and swans tend to begin their molt earlier than breeders and in fact band together for a molt migration, flying to some secluded spot where they shed their feathers en masse, in relative safety. Breeding pairs usually don't molt at the same time. Females generally begin about two weeks after the brood hatches; males don't start losing feathers until several weeks later, about the time their mates are regaining the ability to fly. At least one parent is therefore always fully capable of defending the brood. By the time the young are completely fledged, which from hatching takes forty-nine to sixty-three days for snow geese and sixty-three to eighty-six days for the larger races of Canadas, the male's flight feathers are grown in, and the whole family can take off together.

As shortening days and lowering angles of sunlight signal the onset of autumn, northern-nesting families and flocks gather at traditional staging areas, massing together by the thousands, building up their reserves of energy until something—a front of heavy weather or just an inborn sense that it's time to go—sends them up to form their vees and lines and head south once again. They may reach their wintering grounds in a virtually nonstop marathon flight or take their time about it, depending upc weather and habitat along the way. At a number of places alc the major flyways traditional stopover points are managed as s or federal refuges, and there during the peak of migration you can see some of the grandest spectacles nature has to offer.

For several years I lived within a half-hour's drive of one such place, a spot particularly favored by snow geese. If you've never seen it, believe me that 300,000 geese coming off a single lake in the first faint light of dawn is an impressive sight. The clamor of their voices carries for miles. Up close, combined with the thunderous roar of wings, it's nearly deafening. Even farther off, when you're crouched in a cold, muddy pit-blind surrounded by decoys, not all the shivers chasing up and down your back are caused by the north wind.

Similar shivers accrue to a flock of snows sparkling high and icy-white against a bluebird sky, catching light from a sun still hidden below the eastern rim of the world. And to a night-time flock winging overhead, luminescent with moonlight.

Large flocks of snow geese seldom visit the center of Missouri, where I've lived the past dozen years. This country belongs instead to Canadas, especially to a local flock of giants that stay around through all but the coldest weeks of the year. As I write this, it's early August and I have once again watched our resident pair bring off a brood at the ponds below the house. The three goslings that survived from a clutch of four eggs are now adult-sized and flying.

This morning, as on most mornings, they took off just at dawn, wheeled up from the little lake, circled once, and lined out over the trees, slanting toward the sunrise and whatever feeding ground they'll visit today. Their voices were crisp in the morning quiet, their presence a chantry for early-waking souls, endowing a vision of space and time. What a sere and shrunken place the world would seem without voyagers bringing tales from beyond the wind.

# 29

# Great Northern Diver

As if to give us yet another reason for being on the water at that time of night, the moon came up behind the jagged skyline of spruce and seemed to pose there. We both stopped fishing, lit our pipes, and followed the Night Lady's progress, watching her face grow cool and pale and distant against a field of prickly-looking stars. In another corner of the northern-Minnesota sky the aurora went on rippling like a curtain breezed aside at some great window.

From somewhere down the moonlight path, now diffused more faintly across the water, two loons struck up a conversation, a pair of misadjusted clarinets discussing their states of disrepair.

Ted leaned back in his swivel-seat, blew a plume of pipesmoke toward the sky, and began reciting A. A. Milne, the poem that begins, "James James Morrison Morrison Weatherby George Dupree..."

I picked up the next line: "'Took great care of his mother though he was only three.'"

"Loo-OOOOOOh," said one of the loons.

"'James James said to his mother, "Mother," he said, said he,'" said Ted.

"Loo-lOOOO-loooo," said the other loon.

"'You must never go down to the end of the town if you don't go down with me.'" We finished the first stanza in unison and took off on the second, just as the loons started their own duet.

The next few minutes must have been unnerving for any insomniac cabin-dwellers on Round Lake, convinced that the place had been suddenly invaded by legions of the profoundly disturbed. A loon chorus is one thing, but a children's poem as accompaniment, delivered in the best declamatory style, probably isn't something you hear there every night, or even every full moon. It was by any definition lunacy of a high order.

Actually *loon* is not derived from the same linguistic source as *lunatic,* though it ought to be. It comes instead, the etymologists insist, from Scandinavian words meaning "clumsy" and is close kin to the English verb *to lumber.* "Awkward" fairly enough describes a loon on land, but even if the bird has no linguistic claim on lunacy, it's impossible to hear one chortling through the darkness of a north-woods night and not imagine a madness so exquisite as to approach the sublime.

Such duality runs deep in the nature of loons, even to the point of paradox. They are powerful in flight yet scarcely able to take off and are more at home in water than anywhere else. They spend their summers on freshwater lakes and their winters on saltwater coasts. During the migrations in spring and fall you might find loons almost anywhere in the United States, yet they literally are the signature of northern wilderness. Any way you look at them loons are extraordinary birds.

Loons belong to both an order, Gaviiformes, and a family, Gaviidae, all their own and to a single genus, *Gavia.* Ornithologists disagree on exactly how many species and subspecies there are, but majority opinion at the moment recognizes four main species, all of which occur only in the Northern Hemisphere. The yellow-billed loon is mainly a Eurasian bird, but it also lives along the Alaskan coast. Both the arctic and red-throated loons are circumpolar. The common loon, *Gavia immer*, belongs primarily to North America and breeds all across Canada and the northern edge of the United States; it winters along both coasts from Alaska to Baja California and from the Maritimes south to Florida and westward along the Gulf Coast to the tip of Texas.

In plumage and size the four species fall into two distinct groups. Common and yellow-billed loons have green heads and throat bands and come similarly dressed in shades of gray lined and spotted with white. The yellow-billed is the largest of all loons, measuring two feet or more from beak to tail and five feet from wingtip to wingtip. Common loons are only slightly smaller.

Arctic and red-throated loons are noticeably smaller than the others, are gray-headed, sport vertical stripes of black and white on their necks, and wear throat patches, also vertical, that are either blackish or reddish according to the species. All loons, regardless of species, look much the same in winter—white underneath and brownish-gray on top—and both sexes are identical year-round.

Physical structure and behavior, too, show few differences from one species to another, for all loons are wonderfully well adapted to a lifestyle and a livelihood centered on water. In Europe and in older American scientific literature loons are commonly called "divers" (hence the old name "great northern diver" for the common loon), and diving is their specialty.

They dive to feed, since fish make up the greatest portion of their diet, and dive to escape from danger. Sometimes it seems they do it simply because they can. Physically a loon is a nearly perfect mechanism for underwater work. Most birds' bones are hollow for light weight; a loon's are nearly solid and of about the same specific gravity as water. This lends two advantages. The lack of buoyancy requires less energy to submerge and stay underwater, and the solid bones make a particularly strong framework, which is useful in the dense, high-pressure medium of water. Diving loons have been known to reach depths of thirty yards or more.

Powerful legs and big, fully webbed feet provide the main source of thrust, and they're set far back on the body for maximum leverage and a streamlined shape. Loons usually propel themselves underwater by alternate leg kicks while holding their wings tightly against their bodies. To put on a burst of speed, however, they kick both legs simultaneously and row with their wings as well, froglike and birdlike all at once. No fish is fast or nimble enough to outstrip or outmaneuver a loon with its heart set on having a meal.

All loons dive, but the technique seems to vary somewhat among the different species. Arctic and red-throated loons begin by rearing up and arcing their bodies forward in a beak-first plunge. The others, the master divers, simply exhale, squeeze tightly with their wings to expel air from their feathers, and disappear underwater with scarcely a ripple left behind.

When a feeding loon reappears after a dive it comes fully to the surface, usually holding a fish, which it will swallow headfirst. (Despite their sharply pointed bills loons do not spear their food as some early naturalists supposed.) But if the bird has dived to escape some real or imagined danger, it might well surface only far enough to periscope its head above water and check things out. A truly frightened loon often heads underwater for the nearest reedbed and stays there submerged with only its bill showing, exposing no more of itself than it must in order to breathe. A loon can hide that way, snorkling comfortably, for a long time.

Such splendid adaptations inevitably require some trade off, and the equipment that serves so well in water is a serious handicap on land and in the air. For one thing, a loon's legs are set so far from its body's center of balance that it literally cannot stand up; like a seal, it has to make its way on land by wriggling, lurching, sliding on its breast, and in general flinging itself along.

Getting airborne is similarly difficult. Long, narrow wings provide little lift for a bulky body. Loons can't take off from land at all and without about a quarter-mile of runway space, they can't take off from water, either. Even then, it's a lumbersome process. The bird runs across the water, heading into the wind, feet and wings thrashing the surface until it works up sufficient speed for liftoff. I never see a departing loon without feeling a bit surprised when it finally gets on the wing and can't help wondering if the destination is worth all the trouble.

Once aloft, though, loons fly with astonishing power, sleek projectiles moving along on steadily beating wings, heads held slightly lower than their bodies, big feet ruddering out behind. Cruising speed, confirmed by radar, is about seventy-five miles per hour, and a loon in a hurry can top ninety in short bursts. Old-time wildfowlers gunning migrating loons along the Atlantic coast often wrote of the difficulty involved in intercepting them with charges of shot, and a few were moved to remark that a falling loon—roughly fifteen pounds of dead weight descending at about a mile a minute—could crash right through the bottom of a shooting dory.

Even unshot, a landing loon is not a graceful sight. Since their wing surface is no match for the momentum of their weight they can't brake their descent and hover down as ducks and geese can. Instead loons have to circle lower and lower to the water and finally plow in with great splash and spray, like bombers landing under full throttle. All told they need about as much space to land as they do to take off.

Such ponderous goings and comings are difficult enough under ideal conditions, and it doesn't take much adversity to be too much. Birds that fail to leave north-country lakes before ice encroaches their open water find themselves in a deathtrap, as do those that land on small ponds and potholes anywhere. Migrating loons sometimes mistake the sheen of rain-drenched highways for open water, and even those that don't disable themselves with the shock of power-landing on concrete end up stranded. Wet roads occasionally fool ducks, too, but ducks can easily take off again. Loons cannot.

The birds that survive the southward trip spend the winter in the oceans along the eastern and Gulf coasts, inconspicuous in their drab winter plumage. Toward the end of March, dressed in a fine suit of full colors, they start moving inland and are well into the southern half of their breeding grounds by the beginning of May.

Common loons do not breed until they're nearly three years old, but once they choose mates the couples probably remain together for life. Established pairs seem to prefer nesting on the same lakes year after year.

They also prefer nesting close to water for obvious reasons, either on the margin of land or on a floating mass of reeds, and how much work they put into a nest seems to depend upon where it is. A land-nest may be no more than a shallow depression about two feet across with a few scattered reed-stems for lining. Floating nests are more elaborately built.

Two eggs make up a typical clutch although one often is infertile. Both adults perform the chores of incubation, which takes twenty-five to thirty days. Hatchling loons are ptilopaedic, precocial, and nidifugous—which means they emerge from the egg dressed in down, are active from their first moments of life, and leave the nest immediately. They take to the water right away, but their capacity for anything more than feeble paddling naturally takes a while to develop, so in the nonce they often ride about on their parents' backs. After about two weeks they begin to grow a second coat of down, this one gray, to replace their blackish natal garb.

Even if both eggs hatch, loons seldom finish a summer with more than one chick. Hatchlings fight with each other until one establishes dominance and thereafter gets more food and better care from the parents. The other is likely to die within a few weeks. Often enough a breeding pair ends up with no young at all, thanks to the attentions of muskellunge, pike, pickerel, bass, or some other predatory fish.

Given a bit of luck and about two months' time, the new generation greets the last days of summer wearing a full set of nondescript juvenal feathers and able to fly. For some reason young-of-the-year leave the north country as much as three weeks ahead of the older ones and tend to choose wintering grounds considerably farther south.

All across their range, summer or winter, in fresh water or salt, loons of every age suffer a variety of hardships at the hands of man. They are highly vulnerable to oil spills along seacoasts. In historical times loons were known to breed as far south as Pennsylvania, Iowa, and California, but habitat lost to wetland drainage or lakeshore development has pushed them hundreds of miles farther north. Even where good habitat is available humans often disrupt their breeding and nesting. Loons are fairly persistent nesters and certainly are conscientious parents, but too-close attention from curious, ignorant, or simply careless boaters can drive them off, leaving eggs or chicks exposed to harm.

Man-made reservoirs have to some extent mitigated losses of natural habitat, but if water levels fluctuate too widely as they sometimes do in power-generating reservoirs, loons might abandon a nest already built or simply refuse to nest at all. Because they hunt by sight the birds also are unhappily affected by dredging, shoreline construction, overintensive recreation, and erosion runoff—anything that clouds their water.

The most serious threat to loons at the moment appears to be mercury contamination, either from outright dumping or as a consequence of acid rain, which liberates naturally occurring mercury from soil and stone. This inevitably ends up in water, where it accumulates in the bodies of fish and other aquatic animals and works its way up the food chain. Over the past several years ornithologists have noted massive die-offs among loons arriving on their wintering grounds exhausted, emaciated, their nervous systems out of whack to the point that they're unable to dive—all of which is symptomatic of heavy-metal poisoning. Evidence and contamination continue to accumulate while governments and industries steadfastly avoid the problem or whine over the cost of solving it.

To many, perhaps to most, a loon is only a ghostly sound in the night, and indeed its voice is its signature. Calls vary somewhat from one species to another, but calls almost invariably are the sources of loons' local names. Native peoples of Canada and Alaska, for instance, know the loons as *googara, kaksau, kashgat,* and other names derived from the sounds they make.

The best-known calls naturally belong to the most familiar loon, and while it lacks the sophisticated repertoire that songbirds boast, *Gavia immer* is no vocalistic slouch. The common

loon has four basic calls—officially titled wail, tremolo, yodel, and hoot—and delivers each with some variations. Although we always run the risk of oversimplifying such things, it does seem that each call means something different.

◆

The wail, a long-drawn sound that Thoreau called "looning," appears to express some desire for interaction, a wish to establish contact with mate or young. In the simplest form it is a single tone rendered at about E-flat above middle-C. One variation of it begins at that pitch and then rises about three steps; the bird might end it there or drop back to the original pitch. Another sort of wail covers the same first two notes but ends on a third note about three-quarters of an octave higher yet.

The tremolo is the classic loon call and the only one the birds utter while in flight. It's a giggling, rapidly modulated laugh that from a human source would be, as John McPhee puts it, "the laugh of the deeply insane." From a loon it seems to signal alarm. Like the wail, the tremolo may be a one-, two-, or three-note call ascending in pitch until it sounds like madness incarnate.

Only male loons yodel, or so researchers believe, and the call appears to express aggression or territorial defense. It's a complex, rising series of tones that ends in a multi-note phrase, a little figure the bird might repeat several times depending upon how upset he happens to be. Researchers also have discovered that every male's yodel is individual and distinctive. Since capturing and banding loons is enormously difficult, voice printing may prove useful as a way to keep track of at least the male birds in certain populations.

While the other three calls usually are delivered at impressive volume, the hoot is fetchingly gentle, and loons use it almost conversationally among family groups. It comprises a short, single note, and through different inflections it can sound like a yelp, a whine, or a question.

Loons often combine their calls, possibly for more sophisticated communication, and the constructions seem to be governed by a sort of grammar. For example, when a tremolo is combined with a wail or a yodel, the tremolo always comes first. Interpretation is highly speculative, of course, but meanings of the combinations seem consistent with the meanings of the component parts. A parent loon frightened away from its chicks often calls out a tremolo-wail, which under the circumstances could reasonably be taken to express mixed feelings of anxiety and a wish to reestablish contact.

Of course we'll never know for sure what they're saying to one another, and it hardly matters anyway, so long as the loons know. The world needs a few mysteries left unsolved, just as the north country needs the loon in order to keep some measure of its wildness intact. Like pandas and elephants and mallard ducks, loons capture something in humans, something as elusive as it is important. In the lake country of Minnesota, where the common loon enjoys distinction as the official state bird, you can find whole shops dedicated to the great northern diver's motif, his image on everything from T-shirts to tote bags, doormats to dinnerware, his voice preserved on record and tape.

As if in reflection of the bird's own dualities, loon songs pull us two ways at once. Haunting, they invite us to think about the nature of things, to wonder if the role we've chosen is a rightful one. At other times, they seem to suggest that taking ourselves less seriously might have some value, that grown men might somehow benefit from shouting children's poems on moonlit nights.

# 30

## Angler on Stilts

Even in such a temperate clime as the south-Texas Gulf Coast, a chilly January day on the heels of a blue norther is probably not the best time to be ramming across open water in an airboat. The wind at that speed is a merciless thief, invading coat sleeves and collar and weaseling its way under your hat to steal body heat. But for the muffs you have to wear against the roar of the engine, it would take the hat altogether, and probably your ears as well. Tinted glasses are okay for the glare, but they aren't worth much as a windshield. Aransas Bay is less than three hundred miles from the tropic of Cancer; I've been warmer three hundred miles from the Arctic Circle.

But I had never been on Aransas Bay before (which to me is adequate reason to go almost anywhere, any time), and Herb Booth wanted to show me firsthand the shallow-water flats he paints so beautifully, so there we were—two graybeards who ought to know better, a stoic pilot who plies these waters every day of the year, and the persistent spectre of hypothermia—all having a grand time looping around the bay in a swirl of ducks and shorebirds.

To a lifelong inlander the margin of the sea is an exotic place, a seam at the edge of a vast mystery. Even inside the shelter of the barrier islands it offers glimpses of life lived in scarcely imaginable ways—like the pod of porpoises breaching almost at arm's length next to the boat, as if to have a look at this hideously noisy thing skimming the roof of their world.

And yet against such strangeness lie elements of familiar charm. Again and again as we neared the shore or swung around some grassy shoal, we flushed great blue herons one after another, dozens of them, lifting crook-necked on huge wings, legs dangling clumsily, only to settle down a few yards away and resume their patient vigil for a passing meal.

Great blue herons always make me feel at home, whether I find them in the Pacific Northwest or northern Minnesota, in Nova Scotia or Maryland, South Carolina, Florida, Louisiana, Texas, Mexico, Brazil, or somewhere between. I seem to end up near water no matter where I go, and except for only a few parts of the country, where there's water there's likely to be a blue heron as well, familiar as an old friend.

And they are old friends, partly because my father liked them and invariably pointed out every one we saw together when I was a kid, even the ones I saw first. Herons were a great game for Dad. At the sight of one at the edge of a river or lake, he'd freeze in the instant, caution me not to move either, put down his fishing rod, slowly reach for the oars and try to see how close we could get. We didn't do too badly at our heron-stalking, actually, but you can sooner sneak up on a wild turkey or a will-o'-the-wisp.

What's kept me fascinated is their sheer wildness. The tamest stock-pond in farmdom becomes a little wilderness when there's a great blue heron in it. Old wildfowlers who know the tricks often place a blue heron facsimile near their spread of decoys, knowing that its presence will lull some of the wary edge from Canada geese. There aren't many birds more intelligent than a Canada goose and few whose senses are sharper all-around—but yet the wisest old gander often seems willing to take security cues from a heron.

The great blue occupies a broad range and in most of it

certainly is the most conspicuous of herons. But he belongs to a family both large and cosmopolitan—fifteen genera comprising sixty-four species, scattered through all the continents of the world, and on New Zealand, Madagascar, and a host of other islands. Together they make up the family Ardeidae. Outside the immediate group, their nearest relatives are wood storks, ibises, spoonbills, and flamingos.

The herons are a diverse group physically. Some are tall, long-necked and slender, others small, short-necked, and chunky-bodied. Some ornithologists prefer to think of them in terms of three tribes segregated by behavior rather than shape or size; thus organized, they become day herons, night herons, and tiger herons.

No tiger herons live in North America, but we have an even dozen species from the other two tribes, only half of which we commonly call herons—great blue, little blue, Louisiana heron, green heron, black-crowned and yellow-crowned night herons. We know the others either as egrets or bitterns.

None is particularly colorful. Great, snowy, and cattle egrets are pure white, as are immature little blue herons and some reddish egrets. (I know, a white reddish egret sounds like a contradiction, but there's a good explanation; it's called "ornithology.") A subspecies of the great blue heron, local to southern Florida and the Caribbean, also is white and called, mercifully, great white heron.

Some others—the reddish egret, yellow-crowned night heron, great blue, little blue, green, and Louisiana herons—are predominantly blue-gray with white, black, and rufous patches. The American bittern wears a cryptic dress of brown and white. Only the black-crowned night heron and least bittern own anything like dramatic markings, and they're pretty drab, even at that.

It may not be gaudy, but heron plumage is highly useful and includes some specialized feathers that other birds lack. Powder downs are modified down feathers that gradually disintegrate to produce a talclike substance herons use in preening to remove fish slime and oil from the rest of their feathers. Every member of the family has at least four patches of powder downs, one pair on the breast and another on the rump. Some species grow a third pair high on their inner thighs, and yet others have tracts of them on their backs. Powder from the black-crowned night heron's feathers is slightly luminous.

Both of the typical heron body-shapes are well represented among the North American birds. The bitterns, night herons, and green heron are plump-bodied with relatively short necks and legs. The others are long-necked and long-legged.

◆

With one exception, all of our herons are wading birds and therefore hang out near water. The Louisiana heron and reddish egret mainly prefer saltwater habitats, while the little blue heron and least bittern mainly prefer fresh. The rest don't seem to care and are equally at home around lakes, rivers, ponds, streams, marshes, swamps, shores, and tidal flats. The exception is the cattle egret, a terrestrial, insect-eating, Old World species that made its way from Africa to South America and thence to the United States by way of the West Indies and Florida. It has now spread all the way to southern Canada, and in some places the populations are quite large. True to the name, cattle egrets spend much of their time foraging in pastures among the grazing kine, and you'll often see them perched on some cow's back, waiting to snatch the bugs it stirs up. They do the same in their native Africa with any large animal that will tolerate them, including elephants and Cape buffalo.

Other herons certainly won't pass up the odd grasshopper, dragonfly, or moth, but insects are seldom their first choice. Fish make up as much as 90 percent of their diets, supplemented with frogs, crawfish, salamanders, small snakes, and other aquatic fauna. Some feed by day, others by night. The larger herons occasionally knock off small rodents and shorebirds. The longer-legged species usually wade for their prey, sometimes waiting patiently for fish to swim past, sometimes stalking slowly, sometimes dashing around in pursuit. Those with shorter legs frequently perch on tree roots, low-hanging branches, or flotsam and dive at fish that happen by.

The African black heron adds an interesting dimension by extending its wings forward to form a canopy over its head while wading. The shade may help lure fish and may also help the bird's vision by cutting glare from the water's surface. On this continent, the reddish egret also fishes with its wings raised but not in quite the same way.

Although outnumbered by other species in some parts of the country, the great blue is the quintessential heron and one of the most distinctive birds in North America. Sandhill and whooping cranes have a similar shape, but cranes fly with their necks straight out while herons and egrets loop theirs into a flat S-shape, drawing their heads nearly to their shoulders. Otherwise it's hard to mistake a heron for anything else. Or as Hamlet puts it, fencing words with Rosencrantz and Guildenstern, "I am but mad north-northwest. When the wind is southerly I know a hawk from a hand-saw"—Shakespeare's quibble on *hernshaw,* an old British name for the gray heron.

Perhaps size has something to do with it. The typical adult great blue heron measures nearly forty inches from beak to tail and has a wingspread of about six feet. Among American birds only the cranes and the flamingo stand taller, and only the cranes, vultures, and eagles can match the wingspan.

But for all his impressive dimensions, Old Blue is a flyweight of about seven pounds on the average, seldom as much as eight—truly, as my old friend naturalist-writer John Madson describes him, "mostly bones and underwear." Denuded of all its feathers—a thoroughly unappetizing sight I've seen once and hope not to see again—a blue heron is grotesquely reptilian. It looks like a scrawny chicken with a snake grafted onto one end and a pair of stilts onto the other.

Fully dressed it's another story. Not even the kindest observer would call a great blue heron swan-graceful, but it still manages to carry off an air of gawky dignity—about what you'd get if Ichabod Crane had been an undertaker instead of a schoolteacher. And to watch one stalking fish is to have a glimpse of natural poetry.

No cat is a stealthier hunter. Herons don't hide in ambush, exactly, but they don't need to. Underwater, its sticklike legs appear to be just that, or reeds perhaps, homely and unalarming as the bird makes its patient way step by slow step. Often it pauses with one foot raised, balancing effortlessly, its head slightly cocked, neck stretched for a good view. With prey in range it fixes a careful bead, draws back its head and strikes, quick as a snake or an assassin flashing a dagger. Through the millennia the vertebrae in its neck have grown specially adapted both to perform the motion and to withstand the whiplike shock.

Although the great blue heron's bill is six inches long and sharp-tipped, it's a grasping tool rather than a spearhead. In all the hours I've spent watching scores of herons fish, I can't recall more than a few instances when one actually impaled anything. Popular literature often refers to herons "spearing" fish as if that were their intention, but I have a notion these harpoonings are accidental, that they happen when the bird misjudges distance and stabs its prey before it can open its bill. When the same structure serves the dual functions of catching prey and eating it, too, deliberate spearing doesn't make much sense. If it impales something, the bird has to wade ashore or onto a sandbar and rake the thing off with its foot before eating; then it has to wade back to the fishing ground and begin stalking all over again. There's no such thing as a fat heron, but they'd be skinnier still if they were no more efficient than that.

Instead, a heron comes up from a successful strike with a fish or frog scissored in its bill. It turns the prey headfirst with a couple of deft flicks and swallows it. If you're close enough or watching through powerful glasses you can see the bulge slide down the slender neck. Since a great blue is capable of taking fish that weigh a pound or more, accidental chokings are not unheard of.

Although you'll sometimes find large numbers of them feeding in a relatively small area, blue herons do not live in flocks as some other herons do. All species are more or less gregarious during the nesting season, however, and a heronry full of great blues is something to behold.

So are the courtship rituals that are preamble to the whole business. Some species grow long, diaphanous plumes for the occasion. These aigrettes are the source of the name *egret* and once were the source of near-extinction as well. At the turn of the century aigrettes were highly prized as decoration for women's hats, and plume-hunters killed birds by the thousand to supply the millinery trade. Unfortunately egrets grow these feathers only during the mating season, so the death of plume-bearing adults also meant the demise of helpless nestlings. Thus bedeviled by the slaughter of two generations at once, some species almost disappeared. None actually did go extinct before they were afforded legal protection, but it was a near thing.

(Plumage isn't the only reason why herons have suffered at the hands of man. In his text for a 1937 edition of Audubon's paintings, William Vogt reports that sponge-fishers in the Florida Keys nearly wiped out the great white heron by taking its young for food. I suppose any animal is edible if you're hungry enough, but squab of heron strikes me as about as palatable as vulture fricassee.)

Even those who don't grow fabulous costumes to impress their lady-friends perform some sort of courtship display. The male blue heron executes a nice maneuver in which he points his bill straight up, neck fully extended, and bends his head backwards to show off both the fluffy, beardlike feathers on his upper breast and his long black topknot. Courting is a fairly noisy affair as well, although their repertoire of croaks, creaks, and raspy honking could only sound endearing to another heron. Nonetheless it all seems to work, and once a pair takes up with one another, they will stay together until nesting ends.

Great blue herons use the same rookery year after year— a riverbottom grove of sycamores, cottonwoods, soft maples, or whatever else grows tall and strong enough to support their enormous nests. Male and female alike share in building and refurbishing these haybale-sized structures, made mostly of sticks anchored to the treetops so skillfully that they can withstand almost any weather short of a hurricane. The number of nests naturally depends upon how many birds make up the local colony, but several dozen is the average.

In due course the females lay three to five greenish blue eggs apiece, and the adults take turns incubating them for about four weeks. The young are nidicolous, which means they stay in the nest or nearby until they're old enough to go out foraging on their own. The parents feed them in the meantime, and that, along with a few other components, makes a heronry a place to test any birdwatcher's mettle.

For one thing it's almost sure to be located atop some dense, airless, humid riverine jungle full of poison ivy, stinging nettle, and mosquitoes. The fetid stink of mud, decaying vegetation, and pools of stagnant water is bad enough, but then there's *eau de heron* itself. Since the adults eat fish, they feed the kids fish as well, which they regurgitate and extrude through their bills for the young to gobble up. Not all of this stuff gets down those hungry little maws, but what does eventually ends up on the ground or splattered on the trees. Add in a substantial contribution from the adults, who can void excrement in quantities to impress a Guernsey bull, and the atmosphere is enough to bring tears to the eyes of a skunk.

I don't mean to dwell on the subject, but there's more to heron droppings than just the stench. It also is highly acidic stuff, so by the time the young are fledged and the colony abandons the site, the nest trees wear a ragged mantle of withered leaves, dead twigs, and branch-ends.

Perhaps it's just as well that these are unpleasant places. The congregations are every bit as shy as the individuals, and the less the rookeries are disturbed the better their success in bringing forth new generations.

Still, a heronry is not without its charms, like the one I found a few springs ago along the Niangua River in south Missouri. It's a quiet, lovely stream, the Niangua—or at least it is until the summertime hordes of floaters show up and begin treating it like an amusement park. But this was May, still too early for the aluminum armada, and our little party had the river all to ourselves.

Those who have spent much time with me on the rivers are fond of saying that I could find some way to lag behind while going over a waterfall. This is probably true and troubles me not at all; a river offers too many things to see and feel to be in any hurry. So I was poking along as usual when I heard the unmistakable croak of herons and turned my canoe into the mouth of the next feeder stream that came up on the left. The sounds grew louder as the stream narrowed and shallowed until finally, using the paddle as a pushpole, I reached the end in a muddy backwater scarcely wider than the length of the boat. And there in the treetops all around were the nests, looming dark and bulky against the sun-dappled canopy overhead.

Surprisingly, the birds showed no sign of knowing I was there—thanks, I suspect, to the combination of the extreme quiet of my dear old Mad River canoe, a slow approach screened by a thick understory of leaves, and the fact that at this time of day half the colony was out fishing anyway. It didn't even smell as rank as most heronries. So I eased the boat around for the best view, lay back against my seat cushion, and settled in to watch—hoping that as one final stroke of luck no heron would decide to empty its bowels in a strafing run over the pool.

It was a scene dredged up from the basement of time. The birds were still incubating so there was no hungry clamor—in fact not much sound at all apart from the drowsy hum of insects, and a few rusty clucks and occasional raspy honking as the fishers returned one by one to change shifts, wheeling smoothly above the trees, legs trailing. Their enormous wings cast drifting shadows, blotting out the myriad pinpoints and patches of blue where sunlight filtered through the thick tracery of leaves.

After each returning bird sailed a couple of circles overhead, a huge pair of wings would unfold from one of the nests and stroke slowly as the brooder lifted off to cut a circle or two of its own, exchanging what I presume were a few civilities with its mate. Then one would head off while the other homed in on the nest, reversing its wings to cut airspeed and lowering its gangly landing gear, coming to rest with a delicacy astonishing in a bird so large. It would stand for a moment folding its wings and then disappear behind the shadow of the nest. This was a small colony—a dozen or fifteen nests as best I could tell—but there was enough activity to keep me mesmerized for an hour or more.

The longer I lay there the better I could imagine a rookery of pterodactyls gliding above a great fern forest. Easing back down to the river, I would not have been surprised to meet a dinosaur coming the other way.

Back on the sunny river I paddled on, for once oblivious to everything except the slowly fading vision of a world out of time. At the next bend a green heron spooked out of the streamside bushes as green herons do, beating its two-foot wings in deep strokes and uttering its little call that sounds like *skew* spoken on a high, descending note.

And as green herons also do, it landed in a tree a hundred yards farther on, flushing again as I approached. We went on that way for a half mile or more, the bird going from tree to tree trailing its bright-orange legs, staying in sight as if to show the way downstream.

◆

# About *Wildlife Art* Magazine

W*ildlife Art* magazine started as a newsletter and has grown into the world's largest and most widely recognized journal for information about art and artists depicting the natural world. This award-winning publication is distributed to over seventy countries with loyal subscibers who love the outdoors, animals, and art.

Many thanks to the following artists whose work graces the pages of this book:

Paul J. Alico, Hamburg, N.Y.; Karen Boylan, Bozeman, Mont.; Karen Ray Brower, Butte, Mont.; Dennis L. Burkhart, Wrightsville, Pa.; Deborah Camero, Warminster, Pa.; George Dinan, Vancouver, Wash.; Michael Dustin, Meridian, Idaho; Michael Gelina, Williamson, N.Y.; James Hautman, Plymouth, Minn.; Diane Iverson, Huntington Beach, Calif.; Chris Jacobson, Plover, Wis.; Skeeter Leard, Cerrillos, N.M.; Lee LeBlanc (1913–1988); James Long, Modesto, Calif.; Donald McMichael, North Bend, Ore.; Judith Angell Meyer, Greeley, Ohio; Wanda Mumm, Woodbury, Minn.; Dee Dee Murry, Centralia, Wash.; Dominique Paulus, Jefferson, N.H.; Paul Rawhouser, Big Piney, Wyo.; Martiena Richter, Rochester, Minn.; Elizabeth Ruhl, Cody, Wyo.; Bart Rulon, Greenbank, Wash.; Richard Salvucci; Brighton, Mass.; David Vollbracht, Medicine Lodge, Kans.; Sandra M. Wiesman-Weiler, Mansfield, Wis.; Walter Wilwerding (1891–1966); Kenneth Wind, Laurens, Iowa; James Zenner, Mosinee,Wis.

# Index